PREDICATIVE ARITHMETIC

by

Edward Nelson

Mathematical Notes 32

Princeton University Press

Princeton, New Jersey

1986

Printed in the United States of America by
Princeton University Press, 41 William Street,
Princeton, New Jersey 08540

ISBN 0-691-08455-6

Library of Congress Cataloging in Publication Data
will be found on the last printed page of this book

To Nancy

I put myself into this work,
so there is also something of you in it.

Acknowledgments

This work was begun in December, 1979, at the Institute for Advanced Study. The next year I lectured on some of this material at Princeton University, and benefited from the comments of David Anderson, Jay Hook, Patrick Smith, Hale Trotter, Brian White, and Mitsuru Yasuhara. I also had helpful conversations with Laurence Kirby and Simon Kochen, and especially with Sam Buss, who found many errors and made many suggestions for improvements. The work was partially supported by the National Science Foundation. I am grateful to these people and institutions.

Table of contents

Chapter 1

The impredicativity of induction

The induction principle is this: if a property holds for 0, and if whenever it holds for a number n it also holds for $n+1$, then the property holds for all numbers. For example, let $\theta(n)$ be the property that there exists a number m such that $2 \cdot m = n \cdot (n+1)$. Then $\theta(0)$ (let $m = 0$). Suppose $2 \cdot m = n \cdot (n+1)$. Then $2 \cdot (m+n+1) = (n+1) \cdot ((n+1)+1)$, and thus if $\theta(n)$ then $\theta(n+1)$. The induction principle allows us to conclude $\theta(n)$ for all numbers n. As a second example, let $\pi(n)$ be the property that there exists a non-zero number m that is divisible by all numbers from 1 to n. Then $\pi(0)$ (let $m = 1$). Suppose m is a non-zero number that is divisible by all numbers from 1 to n. Then $m \cdot (n+1)$ is a non-zero number that is divisible by all numbers from 1 to $n+1$, and thus if $\pi(n)$ then $\pi(n+1)$. The induction principle would allow us to conclude $\pi(n)$ for all numbers n.

The reason for mistrusting the induction principle is that it involves an impredicative concept of number. It is not correct to argue that induction only involves the numbers from 0 to n; the property of n being established may be a formula with bound variables that are thought of as ranging over all numbers. That is, the induction principle assumes that the natural number system is given. A number is conceived to be an object satisfying every inductive formula; for a particular inductive formula, therefore, the bound variables are conceived to range over objects satisfying every inductive formula, including the one in question.

In the first example, at least one can say in advance how big is the number m whose existence is asserted by $\theta(n)$: it is no bigger than $n \cdot (n+1)$. This induction is bounded, and one can hope that a predicative treatment of numbers can be constructed that yields the result $\theta(n)$. In the second example, the number m whose existence is asserted by $\pi(n)$ cannot be bounded in terms of the data of the problem.

It appears to be universally taken for granted by mathematicians, whatever their views on foundational questions may be, that the impredicativity inherent in the induction principle is harmless—that there is a concept of number given in advance of all mathematical constructions, that discourse within the domain of numbers is meaningful. But numbers are symbolic constructions; a construction does not exist until it is made; when something new is made, it is something new and not a selection from a pre-existing collection. There is no map of the world because the world is coming into being.

Let us explore the possibility of developing arithmetic predicatively.

Chapter 2

Logical terminology

I tried several times to write a brief, clear summary of the logical terminology that will be used in this investigation, but it always came out long and muddy. Instead, I refer the reader to the beautiful exposition in Shoenfield's book [Sh], especially the first four chapters. Our logical terminology and notation are those of [Sh] except for some departures and additions that will be indicated.

Lower case italic letters, possibly with 0, 1, ... as a subscript, are variables. The order $a, b, \ldots, z, a_0, b_0, \ldots, z_0, a_1, \ldots$ of the variables is called *alphabetical order*. Roman letters are used as in [Sh] as syntactical variables when talking about expressions.

We define $\mathrm{A_x[a]}$ as follows: if no variable occurring in the term a occurs bound in the formula A, substitute a for each free occurrence of the variable x in A; otherwise let $\mathrm{x}_1, \ldots, \mathrm{x}_\nu$ be in alphabetical order the variables that occur in a and occur bound in A, let $\mathrm{y}_1, \ldots, \mathrm{y}_\nu$ be in alphabetical order the first ν variables distinct from all variables occurring in A or a, substitute y_μ for each bound occurrence of x_μ in A for all μ from 1 to ν, and then substitute a for each free occurrence of x. For example, if A is

$$\forall x \forall y\ (x+y)+z = x+(y+z),$$

then $\mathrm{A}_z[0]$ is $\forall x \forall y\ (x+y)+0 = x+(y+0)$ and $\mathrm{A}_z[x]$ is $\forall a \forall y\ (a+y)+x = a+(y+x)$. We write

$$\mathrm{A}_{\mathrm{x}_1 \ldots \mathrm{x}_\nu}[\mathrm{a}_1 \ldots \mathrm{a}_\nu] \quad \text{for} \quad \mathrm{A}_{\mathrm{x}_1}[\mathrm{a}_1] \ldots_{\mathrm{x}_\nu}[\mathrm{a}_\nu].$$

We make the following abbreviations, in addition to those of [Sh]. If $p_1, \ldots, p_\nu$ are binary predicate symbols, we write

$$a_0 p_1 a_1 \ldots a_{\nu-1} p_\nu a_\nu \quad \text{for} \quad a_0 p_1 a_1 \ \& \ \cdots \ \& \ a_{\nu-1} p_\nu a_\nu,$$

as in $x = y = z$ for $x = y$ & $y = z$. We follow [Sh] is associating $\vee$, &, and $\rightarrow$ from right to left, and letting $\vee$ and & take precedence over $\rightarrow$ and $\leftrightarrow$ in restoring parentheses, so that

$$A_1 \rightarrow \cdots \rightarrow A_{\nu-1} \rightarrow A_\nu$$

is equivalent to

$$A_1 \ \& \ \cdots \ \& \ A_{\nu-1} \rightarrow A_\nu,$$

but we adopt the convention that

$$A_1 \leftrightarrow \cdots \leftrightarrow A_{\nu-1} \leftrightarrow A_\nu$$

is an abbreviation for

$$(A_1 \leftrightarrow A_2) \ \& \ \cdots \ \& \ (A_{\nu-1} \leftrightarrow A_\nu).$$

If A is $A_1 \rightarrow A_2$, we write *hyp*A for A_1 and *con*A for A_2. If A is $A_1 \leftrightarrow A_2$, we write *lhs*A for A_1 and *rhs*A for A_2. We write

$$\exists! x A \quad \text{for} \quad \exists x (A \ \& \ \forall y (A_x[y] \rightarrow y = x)),$$

where y is the first variable in alphabetical order distinct from x and all variables occurring in A. If all bound occurrences of x in A occur in the part $\exists x B$, we write $\mathit{scope}_{\exists x} A$ for B, and similarly with $\exists$ replaced by $\exists!$ or $\forall$.

We write the defining axiom of a function symbol f as

$$f x_1 \ldots x_\nu = y \leftrightarrow D$$

(instead of $y = f x_1 \ldots x_\nu \leftrightarrow D$ as in [Sh,§4.6]). Sometimes we write it in the form

$$f x_1 \ldots x_\nu = y \leftrightarrow A, \quad \text{otherwise} \quad y = e$$

where e is a constant. This is an abbreviation for

$$f x_1 \ldots x_\nu = y \leftrightarrow A \vee (\neg \exists y A \ \& \ y = e).$$

Then the existence condition is trivial, but the uniqueness condition still has to be verified. Also, we sometimes adjoin a new function symbol f by writing

$$\mathrm{f}\mathrm{x}_1 \ldots \mathrm{x}_\nu = \mathrm{a}$$

where a is a term containing no variable other than $\mathrm{x}_1, \ldots, \mathrm{x}_\nu$; the defining axiom is understood to be

$$\mathrm{f}\mathrm{x}_1 \ldots \mathrm{x}_\nu = \mathrm{y} \leftrightarrow \mathrm{y} = \mathrm{a}$$

where y is the first variable in alphabetical order distinct from $\mathrm{x}_1, \ldots, \mathrm{x}_\nu$.

A formula will be called *unary* in case one and only one variable occurs free in it. If C is unary and x is the variable occurring free in it, we write $\mathrm{C}[\mathrm{a}]$ for $\mathrm{C}_\mathrm{x}[\mathrm{a}]$. Let C be a unary formula. Then we write A_C for the formula obtained by replacing each part of A of the form $\exists\mathrm{yB}$ by $\exists\mathrm{y}(\mathrm{C}[\mathrm{y}] \ \& \ \mathrm{B})$. (We follow [Sh] in regarding $\forall\mathrm{y}$ as an abbreviation for $\neg\exists\mathrm{y}\neg$; if one chooses not to eliminate the defined symbol $\forall$, then an equivalent formula is obtained if in addition one replaces each part of A of the form $\forall\mathrm{yB}$ by $\forall\mathrm{y}(\mathrm{C}[\mathrm{y}] \rightarrow \mathrm{B})$.) Let $\mathrm{x}_1, \ldots, \mathrm{x}_\nu$ be in alphabetical order the variables occurring free in A; then we write

$$\mathrm{C}(\mathit{free}\mathrm{A}) \quad \text{for} \quad \mathrm{C}[\mathrm{x}_1] \ \& \ \cdots \ \& \ \mathrm{C}[\mathrm{x}_\nu].$$

If A is closed, then $\mathrm{C}(\mathit{free}\mathrm{A})$ is the empty expression; we make the convention that if A is closed, then all occurrences of "$\mathrm{C}(\mathit{free}\mathrm{A})$" together with attendant logical connectives are to be deleted. (In general, I will not worry about the distinction between use and mention, but in this case I use quotation marks lest the reader be puzzled as to how to go about deleting the empty expression.) We write

$$\mathrm{A}^\mathrm{C} \quad \text{for} \quad \mathrm{C}(\mathit{free}\mathrm{A}) \rightarrow \mathrm{A}_\mathrm{C}$$

and call A^C the *relativization* of A by C.

Let C be a unary formula of a theory T. For f a ν-ary function symbol, we say that C *respects* f in case

$$\vdash_T \mathrm{C}[x_1] \rightarrow \cdots \rightarrow \mathrm{C}[x_\nu] \rightarrow \mathrm{C}[\mathrm{f}x_1 \ldots x_\nu].$$

We say that C *respects* A in case $\vdash_T A^C$. We say that C is *inductive* in case C respects 0 and S (assuming T to contain the constant 0 and the unary function symbol S), so that C is inductive if and only if

$$\vdash_T C[0] \ \& \ (C[x] \to C[Sx])$$

(or, equivalently, if and only if $\vdash_T C[0] \ \& \ \forall x(C[x] \to C[Sx])$). We say that C is *hereditary* in case

$$\vdash_T C[x] \ \& \ y \le x \to C[y]$$

(assuming T to contain the binary predicate symbol $\le$). If C′ is also unary, we say that C′ is *stronger than* C in case

$$\vdash_T C'[x] \to C[x].$$

If x occurs free in A and $y_1, \ldots, y_\nu$ are in alphabetical order the variables distinct from x that occur free in A, we write

$$A_{/x} \quad \text{for} \quad \forall y_1 \cdots \forall y_\nu A,$$

and we say that A in *inductive in* x in case $A_{/x}$ is inductive. If p is a unary predicate sumbol, we say that p is inductive, etc., in case the formula $p(x)$ is inductive, etc.

The definitions in the preceding paragraph are all relative to the theory T. If T is not clear from the context, we add *in* T.

We write

$$ind_x A \quad \text{for} \quad A_x[0] \ \& \ \forall x(A \to A_x[Sx]).$$

An *induction formula* is a formula of the form $ind_x A \to A$.

We write $T[B_1, \ldots, B_\lambda]$ for the theory obtained by adjoining to the language of T all of the new nonlogical symbols in $B_1, \ldots, B_\lambda$ and then adjoining $B_1, \ldots, B_\lambda$ as nonlogical axioms.

Let C be a unary formula of the theory T. We say that C *respects* T in case C respects every function symbol and nonlogical axiom of T and $\vdash_T \exists x C[x]$. This last condition follows from the first if T contains a constant (0-ary function symbol) e, for then $\vdash_T C[e]$. Let C respect T, let U be the extension by definitions of T obtained by adjoining the unary predicate symbol U_I with defining axiom $U_I x \leftrightarrow C[x]$, and let u_I be u for each nonlogical sumbol u of T. Then we have an interpretation (see [Sh,§4.7]) of T in U, which we call the interpretation *associated* with C.

We write (ξ), where ξ is some label, as an abbreviation for the formula with that label, and we do not hesitate to treat it as a formula. For example, the formula with label 3.4 will be found in Chapter 3; it is $x + Sy = S(x+y)$, so $(3.4)_y[0]$ is $x + S0 = S(x+0)$.

When we introduce a hypothesis in a proof we use the word suppose, and the discharge of the hypothesis is indicated by the word thus. (Suppose we adopt this convention (suppose, that is, that in proofs we use these two words (suppose and thus) in the manner indicated, and only thus). Then these two words (suppose and thus) function much as parentheses (suppose playing the role of the left parenthesis, with the role of the right parenthesis being played by thus). We hope to achieve clarity thus.)

As is customary in mathematics, we use letters—for the purposes of this paragraph, a letter is an italic lower case letter possibly with 0, 1, ... as a subscript—sometimes as variables and, in the course of a proof, sometimes as constants, and a single letter may play both roles in a given proof. When we have proved $\exists x_1 \cdots \exists x_\nu A$, we indicate the introduction of constants (see the discussion of special constants in [Sh,§4.2]) by writing

there exist x_1, ..., and x_ν such that A

(with the appropriate change in grammar or punctuation when ν is 1 or 2). We write

let $x_1 = a_1, \ldots, x_\nu = a_\nu$

instead of

there exist x_1, ..., and x_ν such that $x_1 = a_1$ & $\cdots$ & $x_\nu = a_\nu$.

A *numeral* is an expression of the form $S \cdots S0$. We use n, possibly with a subscript, as a syntactical variable for numerals, and $\bar{\nu}$ is an abbreviation for the numeral with ν occurrences of S.

Chapter 3

The axioms of arithmetic

By *Peano Arithmetic* we mean the theory I whose nonlogical symbols are the constant 0, the unary function symbol S (successor), and the binary function symbols $+$ and $\cdot$, and whose nonlogical axioms are

3.1 *Ax.* $\mathrm{S}x \neq 0$,

3.2 *Ax.* $\mathrm{S}x = \mathrm{S}y \rightarrow x = y$,

3.3 *Ax.* $x + 0 = x$,

3.4 *Ax.* $x + \mathrm{S}y = \mathrm{S}(x + y)$,

3.5 *Ax.* $x \cdot 0 = 0$,

3.6 *Ax.* $x \cdot \mathrm{S}y = x \cdot y + x$,

and all induction formulas in the language of I. We have adopted the usual convention that $\cdot$ takes predecence over $+$ in restoring parentheses, so that (3.6) is an abbreviation for $x \cdot \mathrm{S}y = (x \cdot y) + x$.

If we simply drop the induction formulas as axioms, then the resulting theory is too weak to be of much arithmetical interest; a model-theoretic argument to this effect is given by Mostowski, Robinson, and Tarski in [MRT,pp.62-64].

Robinson's theory (see [Ro]) is the theory Q with the language of I whose nonlogical axioms are (3.1)–(3.6) and the following formula (which is a theorem of I since it is easily proved by induction):

R. $x \neq 0 \rightarrow \exists y\ \mathrm{S}y = x$.

This beautiful and much studied theory is in a sense a minimal axiomatization of arithmetic. We will work only in theories that are interpretable in Q.

It will be convenient to reformulate Q as an open theory. The formula

3.7 *Ax.* $\mathrm{P}x = y \leftrightarrow \mathrm{S}y = x \vee (x = 0 \;\&\; y = 0)$

is the defining axiom of a unary function symbol P (predecessor) that can be adjoined to Q; the existence condition holds by (R) and the uniqueness condition holds by (3.2). Let Q_0 be the theory whose nonlogical symbols are those of Q together with P and whose nonlogical axioms are (3.1)–(3.7); then (R) is a theorem, but not an axiom, of Q_0.

To get to the main point more quickly, let us at first adjoin the associative, distributive, and commutative laws:

3.8 *Ax.* $(x + y) + z = x + (y + z)$,

3.9 *Ax.* $x \cdot (y + z) = x \cdot y + x \cdot z$,

3.10 *Ax.* $(x \cdot y) \cdot z = x \cdot (y \cdot z)$,

3.11 *Ax.* $x + y = y + x$,

3.12 *Ax.* $x \cdot y = y \cdot x$.

Let Q_1 be Q_0 with (3.8)–(3.12) as additional axioms. Later we will investigate how to avoid assuming these as axioms.

Chapter 4

Order

The following formula is the defining axiom of a binary predicate symbol that we adjoin to Q_1:

4.1 *Def.* $x \leq y \leftrightarrow \exists z\ x + z = y$.

Call the resulting theory Q_1'. In this chapter we prove a few simple theorems in Q_1'.

4.2 *Thm.* $0 \leq x \leq x \leq \mathrm{S}x$.

Proof. We have $0 + x = x$ by (3.3) and (3.11), so $0 \leq x$. We have $x + 0 = x$ by (3.3), so $x \leq x$, and $x + \mathrm{S}0 = \mathrm{S}(x + 0) = \mathrm{S}x$ by (3.4) and (3.3), so $x \leq \mathrm{S}x$.

4.3 *Thm.* $x \leq 0 \rightarrow x = 0$.

Proof. Suppose $x \leq 0$. Then there exists z such that $x + z = 0$. Suppose $z = 0$. Then $x = 0$ by (3.3) and thus $z = 0 \rightarrow x = 0$, so suppose $z \neq 0$. Then $\mathrm{SP}z = z$ by (3.7), so that $x + \mathrm{SP}z = 0$, and $\mathrm{S}(x + \mathrm{P}z) = 0$ by (3.4), which is impossible by (3.1). Thus $z = 0$ and $x = 0$, and thus (4.3).

4.4 *Thm.* $y \leq \mathrm{S}x \leftrightarrow y \leq x \vee y = \mathrm{S}x$.

Proof. Suppose $y \leq x$. Then there exists z such that $y + z = x$, and by (3.4) we get $y + \mathrm{S}z = \mathrm{S}(y + z) = \mathrm{S}x$, and so $y \leq \mathrm{S}x$. Thus $y \leq x \rightarrow y \leq \mathrm{S}x$. Suppose $y = \mathrm{S}x$. Then $y \leq \mathrm{S}x$ by (4.2) and thus $rhs(4.4) \rightarrow lhs(4.4)$. Conversely, suppose $y \leq \mathrm{S}x$, so that there exists z such that $y + z = \mathrm{S}x$. Then $z = 0 \rightarrow y = \mathrm{S}x$ by (3.3), so suppose $z \neq 0$. Then $\mathrm{SP}z = z$ by (3.7) and so $y + \mathrm{SP}z = \mathrm{S}x$. By (3.4), $\mathrm{S}(y + \mathrm{P}z) = \mathrm{S}x$ and then $y + \mathrm{P}z = x$ by (3.2), so that $y \leq x$. Thus $z \neq 0 \to y \leq x$, and thus (4.4).

4.5 *Thm.* $x \leq y \ \&\ y \leq z \to x \leq z$.

Proof. Suppose *hyp*(4.5). Then there exist u and v such that $x + u = y \ \&\ y + v = z$. Then $z = (x + u) + v = x + (u + v)$ by (3.8), and so $x \leq z$. Thus (4.5).

4.6 *Thm.* $\mathrm{P}x \leq x$.

Proof. We have $\mathrm{P}0 \leq 0$ by (3.7) and (4.2), so suppose $x \neq 0$. Then $x = \mathrm{SP}x = \mathrm{SP}x + 0 = 0 + \mathrm{SP}x = \mathrm{S}(0 + \mathrm{P}x) = \mathrm{S}(\mathrm{P}x + 0) = \mathrm{P}x + \mathrm{S}0$ by (3.7), (3.3), (3.11), and (3.4), so that $\mathrm{P}x \leq x$. Thus (4.6).

4.7 *Thm.* $x \leq y \to z + x \leq z + y$.

Proof. Suppose $x \leq y$. Then there exists u such that $x + u = y$. Therefore $(z + x) + u = z + (x + u) = z + y$ by (3.8), and thus (4.7).

4.8 *Thm.* $x \leq y \to z \cdot x \leq z \cdot y$.

Proof. Suppose $x \leq y$. Then there exists u such that $x + u = y$. Therefore $z \cdot x + z \cdot u = z \cdot (x + u) = z \cdot y$ by (3.9), and thus (4.8).

Chapter 5

Induction by relativization

Let C be a unary formula. We use the following abbreviations:

C^1 for $\forall y(y \leq x \to C[y])$,

C^2 for $\forall y(C^1[y] \to C^1[y + x])$,

C^3 for $\forall y(C^2[y] \to C^2[y \cdot x])$.

Then C^1, C^2, and C^3 are unary formulas with free variable x. We will show that if C is inductive, then C^3 is stronger than C, is hereditary, and is not only inductive but respects P, $+$, $\cdot$, and the defining axiom of $\leq$. This relativization scheme is due to R. Solovay; see [PD] and [Pu].

Metatheorem 5.1 Let T *be an extension of* Q'_1 *(possibly* Q'_1 *itself), and let* C *be a unary formula of* T. *Then the following is a theorem of* T:

$$\begin{aligned}
\text{REL.}\quad & C[0] \ \&\ \forall x(C[x] \to C[Sx]) \ \to \\
& (C^3[x] \to C[x]) \ \& \\
& (C^3[x] \ \&\ u \leq x \to C^3[u]) \ \& \\
& C^3[0] \ \& \\
& (C^3[x] \to C^3[Sx]) \ \& \\
& (C^3[x] \to C^3[Px]) \ \& \\
& (C^3[x_1] \ \&\ C^3[x_2] \to C^3[x_1 + x_2]) \ \& \\
& (C^3[x_1] \ \&\ C^3[x_2] \to C^3[x_1 \cdot x_2]) \ \& \\
& (C^3[x] \ \&\ C^3[y] \to (x \leq y \leftrightarrow \exists z(C^3[z] \ \&\ x + z = y))).
\end{aligned}$$

Demonstration. We prove (REL) in T as follows. Suppose *hyp*(REL). Then we claim:

1. $\mathrm{C}^1[x] \to \mathrm{C}[x]$,
2. $\mathrm{C}^1[0]$,
3. $\mathrm{C}^1[x] \to \mathrm{C}^1[\mathrm{S}x]$,
4. $\mathrm{C}^1[x]\ \&\ u \leq x \to \mathrm{C}^1[u]$;

5. $\mathrm{C}^2[x] \to \mathrm{C}^1[x]$,
6. $\mathrm{C}^2[0]$,
7. $\mathrm{C}^2[x] \to \mathrm{C}^2[\mathrm{S}x]$,
8. $\mathrm{C}^2[x]\ \&\ u \leq x \to \mathrm{C}^2[u]$,
9. $\mathrm{C}^2[x_1]\ \&\ \mathrm{C}^2[x_2] \to \mathrm{C}^2[x_1 + x_2]$;

10. $\mathrm{C}^3[x] \to \mathrm{C}^2[x]$,
11. $\mathrm{C}^3[0]$,
12. $\mathrm{C}^3[x] \to \mathrm{C}^3[\mathrm{S}x]$,
13. $\mathrm{C}^3[x]\ \&\ u \leq x \to \mathrm{C}^3[u]$,
14. $\mathrm{C}^3[x_1]\ \&\ \mathrm{C}^3[x_2] \to \mathrm{C}^3[x_1 + x_2]$,
15. $\mathrm{C}^3[x_1]\ \&\ \mathrm{C}^3[x_2] \to \mathrm{C}^3[x_1 \cdot x_2]$.

Suppose $\mathrm{C}^1[x]$. By (4.2) we have $\mathrm{C}[x]$, and thus (1). Suppose $y \leq 0$. Then $y = 0$ by (4.3), so that $\mathrm{C}[y]$ and thus (2). Suppose $\mathrm{C}^1[x]\ \&\ y \leq \mathrm{S}x$. By (4.4), $y \leq x \vee y = \mathrm{S}x$. But $y \leq x \to \mathrm{C}[y]$, so suppose $y = \mathrm{S}x$. By (1) we have $\mathrm{C}[x]$ and so $\mathrm{C}[\mathrm{S}x]$, i.e. $\mathrm{C}[y]$. Thus $\mathrm{C}[y]$, and thus (3). Suppose $\mathrm{C}^1[x]\ \&\ u \leq x\ \&\ y \leq u$. Then $y \leq x$ by (4.5), and so $\mathrm{C}[y]$. Thus (4).

Suppose $\mathrm{C}^2[x]$. By (2) we have $\mathrm{C}^1[0 + x]$, but $0 + x = x$ by (3.11) and (3.3), so that $\mathrm{C}^1[x]$, and thus (5). We have (6) by (3.3). Suppose $\mathrm{C}^2[x]\ \&\ \mathrm{C}^1[y]$. Then $\mathrm{C}^1[y + x]$, and so $\mathrm{C}^1[\mathrm{S}(y + x)]$ by (3). But $\mathrm{S}(y + x) = y + \mathrm{S}x$ by (3.4), and so $\mathrm{C}^1[y + \mathrm{S}x]$. Thus (7). Suppose $\mathrm{C}^2[x]\ \&\ u \leq x\ \&\ \mathrm{C}^1[y]$. Then $y + u \leq y + x$ by (4.7), so that $\mathrm{C}^1[y + u]$ by (4), and thus (8). Suppose $\mathrm{C}^2[x_1]\ \&\ \mathrm{C}^2[x_2]\ \&\ \mathrm{C}^1[y]$. By (3.8), $y + (x_1 + x_2) = (y + x_1) + x_2$, but we have $\mathrm{C}^1[y + x_1]$ and therefore $\mathrm{C}^1[(y + x_1) + x_2]$, i.e. $\mathrm{C}^1[y + (x_1 + x_2)]$. Thus (9).

Suppose $\mathrm{C}^3[x]$. We have $\mathrm{C}^2[\mathrm{S}0]$ by (6) and (7), and hence $\mathrm{C}^2[\mathrm{S}0 \cdot x]$. But $\mathrm{S}0 \cdot x = x \cdot \mathrm{S}0 = x \cdot 0 + x = 0 + x = x + 0 = x$ by (3.12), (3.6), (3.5), (3.11), and (3.3), so that $\mathrm{C}^2[x]$ and thus (10). We have (11) by (3.5) and (6). Suppose $\mathrm{C}^3[x]\ \&\ \mathrm{C}^2[y]$. By (3.6), $y \cdot \mathrm{S}x = y \cdot x + y$. We have $\mathrm{C}^2[y \cdot x]$, so (9) yields $\mathrm{C}^2[y \cdot x + y]$, i.e. $\mathrm{C}^2[y \cdot \mathrm{S}x]$, and thus (12). Suppose

$C^3[x]$ & $u \le x$ & $C^2[y]$. Then $y \cdot u \le y \cdot x$ by (4.8). We have $C^2[y \cdot x]$, so $C^2[y \cdot u]$ by (8), and thus (13). Suppose $C^3[x_1]$ & $C^3[x_2]$ & $C^2[y]$. By (3.9), $y \cdot (x_1 + x_2) = y \cdot x_1 + y \cdot x_2$. We have $C^2[y \cdot x_1]$ and $C^2[y \cdot x_2]$, so by (9) we obtain $C^2[y \cdot x_1 + y \cdot x_2]$, i.e. $C^2[y \cdot (x_1 + x_2)]$, and thus (14). Again suppose $C^3[x_1]$ & $C^3[x_2]$ & $C^2[y]$. We have $C^2[y \cdot x_1]$ and therefore $C^2[(y \cdot x_1) \cdot x_2]$. But $(y \cdot x_1) \cdot x_2 = y \cdot (x_1 \cdot x_2)$ by (3.10), so that $C^2[y \cdot (x_1 \cdot x_2)]$, and thus (15).

By (10), (5), and (1) we have $C^3[x] \to C[x]$. By (4.6) and (13) we have $C^3[x] \to C^3[Px]$. Finally, suppose $C^3[x]$ & $C^3[y]$. It is trivial that $\exists z(C^3[z]$ & $x + z = y) \to x \le y$, so suppose $x \le y$. Then there exists z such that $x + z = y$. By (3.11), $z + x = y$, so $z \le y$ and by (13) we obtain $C^3[z]$. Thus $x \le y \leftrightarrow \exists z(C^3[z]$ & $x + z = y)$, and thus $C^3[x]$ & $C^3[y] \to (x \le y \leftrightarrow \exists z(C^3[z]$ & $x + z = y))$. □

If we regard the concept of number as being in need of clarification, then we can seek to clarify the concept by formalizing it. We can do this, for example, by postulating the simple algorithmic properties (3.1)–(3.12), adjoining defining axioms, and proving theorems, such as the theorems of Q_1' proved in Chapter 4. But this theory is very weak, and in it we can prove very little of what we want to prove about numbers. Let C be an inductive formula; our intuitive feeling is that if x is a number, then $C[x]$ should hold. Now the formula C^3 respects all of the function symbols of Q_1' and the defining axiom of $\le$, by REL. All of the other nonlogical axioms of Q_1' are open, so it automatically respects them as well. In other words, the entire theory Q_1' can be relativized by C^3. We can replace our concept of number (any x) by a more refined concept of number (any x such that $C^3[x]$). We can read $C^3[x]$ as "x is a number" (leaving open the possibility of formalizing an even more refined concept of number at some time in the future). We can ask, *if x is a number does* $C[x]$ *hold*? Since C^3 is stronger than C by REL, the answer is yes, if x is a number then $C[x]$. This satisfies our intuition. But now we can ask, *does the formula* C *hold for numbers*? This is a different question. It means: if in $C^3[x] \to C[x]$ we replace the quantifiers $\exists$y in the formula C, which refer to the domain of discourse before we refined our concept of number, by $\exists$y(C^3[y] & ...), which can be read as "there exists a number y such that ..."; in short, if we relativize C by C^3, is the relativized formula correct? This is the crux. This is the point that impredicative arithmetic takes for granted, by postulating an affirmative answer in the induction principle. In predicative arithmetic we examine the relativized formula to see whether we can prove it.

Let us give an example of an inductive formula C for which we can indeed prove that C holds for numbers. Consider the first example of Chapter 1: Let C be $\exists m\ \mathrm{SS0} \cdot m = n \cdot (n + \mathrm{S0})$. Then C is inductive in Q_1'. Consider its relativization by C^3:

$$\mathrm{C}^3[n] \rightarrow \exists m(\mathrm{C}^3[m]\ \&\ \mathrm{SS0} \cdot m = n \cdot (n + \mathrm{S0})).$$

Suppose $\mathrm{C}^3[n]$. Then $\mathrm{C}[n]$, by REL, so there exists m such that $\mathrm{SS0} \cdot m = n \cdot (n + \mathrm{S0})$. Since $\mathrm{SS0} \cdot m = m + m$, we have $m \leq n \cdot (n + \mathrm{S0})$. But we have $\mathrm{C}^3[0]$, $\mathrm{C}^3[\mathrm{S0}]$, $\mathrm{C}^3[n + \mathrm{S0}]$, $\mathrm{C}^3[n \cdot (n + \mathrm{S0})]$, and $\mathrm{C}^3[m]$, all by REL. Thus the displayed formula holds. It can be read "if n is a number, then there is a number m such that $2 \cdot m = n \cdot (n + 1)$".

If now we let C be

$$\exists m(m \neq 0\ \&\ \forall k(k \neq 0\ \&\ k \leq n \rightarrow \exists j\ k \cdot j = n)),$$

which is the second example of Chapter 1, then C is inductive in Q_1' but there seems to be no way to prove its relativization by C^3. We do not have a way of predicating a concept of number such that whenever n is a number, there exists a non-zero number m that is divisible by all numbers from 1 to n.

Chapter 6

Interpretability in Robinson's theory

For the pleasure of working from minimal assumptions, let us show that we can drop the axioms (3.8)–(3.12). We will extend the relativization scheme of the preceding chapter by building into the construction the necessary associativity, etc., and then use this to show that Q_1 is interpretable in Q_0 (and so in Robinson's theory Q). The reader who wishes to skip this chapter can simply substitute Q_1 for Q in later statements about interpretability in Q.

For this chapter, and this chapter only, we make the abbreviations:

α for $\forall x \forall y\ (x + y) + z = x + (y + z)$,

β for $\forall x \forall y\ (\alpha[x] \to x \cdot (y + z) = x \cdot y + x \cdot z)$,

γ for $\forall x \forall y\ (\alpha[x]\ \&\ \beta[y] \to (x \cdot y) \cdot z = x \cdot (y \cdot z))$,

δ for $\forall y\ (0 + y = z \to y = z)$.

Then the following are theorems of Q_0.

6.1 *Thm.* $\alpha[0]$.

Proof. We have $(x + y) + 0 = x + y$ and $x + (y + 0) = x + y$ by (3.3).

6.2 *Thm.* $\alpha[z] \to \alpha[\mathrm{S}z]$.

Proof. Suppose $(x + y) + z = x + (y + z)$. Then $(x + y) + \mathrm{S}z = \mathrm{S}((x + y) + z) = \mathrm{S}(x + (y + z)) = x + \mathrm{S}(y + z) = x + (y + \mathrm{S}z)$ by (3.4). Thus (6.2).

6.3 *Thm.* $\alpha[z] \to \alpha[\mathrm{P}z]$.

Proof. Suppose $z = 0$. Then $\mathrm{P}z = 0$ by (3.7), and thus $z = 0 \to (6.3)$. Suppose $z \neq 0$ & $(x + y) + z = x + (y + z)$, so that $\mathrm{S}((x + y) + \mathrm{P}z) = (x+y)+\mathrm{SP}z = (x+y)+z = x+(y+z) = x+(y+\mathrm{SP}z) = x+\mathrm{S}(y+\mathrm{P}z) = \mathrm{S}(x + (y + \mathrm{P}z))$ by (3.4) and (3.7). Then $(x + y) + \mathrm{P}z = x + (y + \mathrm{P}z)$ by (3.2). Thus $z \neq 0 \to (6.3)$, and so (6.3).

6.4 *Thm.* $\alpha[z_1]$ & $\alpha[z_2] \to \alpha[z_1 + z_2]$.

Proof. Suppose $\alpha[z_1]$ & $\alpha[z_2]$. Then $(x + y) + (z_1 + z_2) = ((x + y) + z_1) + z_2 = (x + (y + z_1)) + z_2 = x + ((y + z_1) + z_2) = x + (y + (z_1 + z_2))$, and thus (6.4).

6.5 *Thm.* $\beta[0]$.

Proof. We have $x \cdot (y + 0) = x \cdot y$ and $x \cdot y + x \cdot 0 = x \cdot y + 0 = x \cdot y$ by (3.3) and (3.5).

6.6 *Thm.* $\beta[z] \to \beta[\mathrm{S}z]$.

Proof. Suppose $\alpha[x]$ & $x \cdot (y + z) = x \cdot y + x \cdot z$. By (3.4) and (3.6), $x \cdot (y + \mathrm{S}z) = x \cdot \mathrm{S}(y + z) = x \cdot (y + z) + x = (x \cdot y + x \cdot z) + x = x \cdot y + (x \cdot z + x) = x \cdot y + x \cdot \mathrm{S}z$. Thus (6.6).

6.7 *Thm.* $\gamma[0]$.

Proof. We have $(x \cdot y) \cdot 0 = 0$ and $x \cdot (y \cdot 0) = x \cdot 0 = 0$ by (3.5).

6.8 *Thm.* $\gamma[z] \to \gamma[\mathrm{S}z]$.

Proof. Suppose $\alpha[x]$ & $\beta[y]$ & $(x \cdot y) \cdot z = x \cdot (y \cdot z)$. By (3.6), $(x \cdot y) \cdot \mathrm{S}z = (x \cdot y) \cdot z + x \cdot y = x \cdot (y \cdot z) + x \cdot y = x \cdot (y \cdot z + y) = x \cdot (y \cdot \mathrm{S}z)$, and thus (6.8).

6.9 *Thm.* $x + y = 0 \to x = 0$ & $y = 0$.

Proof. Suppose $x + y = 0$, and suppose $y = 0$. Then $x = 0$ by (3.5), and thus $y = 0 \to x = 0$, so suppose $y \neq 0$. Then $\mathrm{S}(x + \mathrm{P}y) = 0$ by (3.4) and (3.7), which is impossible by (3.1), and thus $y = 0$. Thus (6.9).

6.10 *Thm.* $\delta[0]$.

Proof. Suppose $0 + y = 0$. Then $y = 0$ by (6.9), and thus (6.10).

6.11 *Thm.* $\delta[x] \rightarrow \delta[Sx]$.

Proof. Suppose $\delta[x]$ & $0 + y = Sx$. Suppose $y = 0$. Then $0 = Sx$ by (3.3), which is absurd by (3.1), and thus $y \neq 0$. By (3.7), $SPy = y$. Hence $0 + SPy = Sx$, so that $S(0 + Py) = Sx$ by (3.4) and $0 + Py = x$ by (3.2). Since $\delta[x]$, we have $Py = x$, so that $SPy = Sx$, i.e. $y = Sx$, and thus (6.11).

6.12 *Thm.* $\delta[0 + x] \rightarrow x = 0 + x$.

Proof. Suppose $\delta[0 + x]$. Since $0 + x = 0 + x$, we have $x = 0 + x$. Thus (6.12). □

Let C be a unary formula. We make the following abbreviations:

$\hat{C}^0$ for $C[x]$ & $\alpha[x]$ & $\beta[x]$ & $\gamma[x]$ & $\delta[x]$,

$\hat{C}^1$ for $\forall y \forall z\ (y + z = x\ \&\ \alpha[z] \rightarrow \hat{C}^0[y])$,

$\hat{C}^2$ for $\forall y\ (\hat{C}^1[y] \rightarrow \hat{C}^1[y + x])$,

$\hat{C}^3$ for $\hat{C}^2[x]\ \&\ \forall y\ (\hat{C}^2[y] \rightarrow \hat{C}^2[y \cdot x])$.

Metatheorem 6.1 Let T *be an extension of* Q_0 *and let* C *be a unary formula of* T. *Then the following is a theorem of* T:

$$
\begin{aligned}
\widehat{\text{REL}}.\quad & C[0]\ \&\ \forall x(C[x] \rightarrow C[Sx]) \rightarrow \\
& (\hat{C}^3[x] \rightarrow C[x])\ \& \\
& \hat{C}^3[0]\ \& \\
& (\hat{C}^3[x] \rightarrow \hat{C}^3[Sx])\ \& \\
& (\hat{C}^3[x] \rightarrow \hat{C}^3[Px])\ \& \\
& (\hat{C}^3[x_1]\ \&\ \hat{C}^3[x_2] \rightarrow \hat{C}^3[x_1 + x_2])\ \& \\
& (\hat{C}^3[x_1]\ \&\ \hat{C}^3[x_2] \rightarrow \hat{C}^3[x_1 \cdot x_2]).
\end{aligned}
$$

Demonstration. We prove ($\widehat{\text{REL}}$) in T as follows. Suppose *hyp*($\widehat{\text{REL}}$). We state and prove a number of claims.

1. $\hat{C}^0[0]$. This holds by (6.1), (6.5), (6.7), and (6.10).
2. $\hat{C}^0[x] \rightarrow \hat{C}^0[Sx]$, by (6.2), (6.6), (6.8), and (6.11).

3. $\hat{C}^1[x] \to \hat{C}^0[x]$. Suppose $\hat{C}^1[x]$. It follows from (3.3) and (6.1) that $x + 0 = x$ & $\alpha[0]$, and consequently $\hat{C}^0[x]$. Thus (3).

4. $\hat{C}^1[0]$. Suppose $y + z = 0$. Then $y = 0$ by (6.9), and so $\hat{C}^0[y]$ by (1). Thus (4).

5. $\hat{C}^1[x] \to \hat{C}^1[Sx]$. Suppose $\hat{C}^1[x]$ & $y + z = Sx$ & $\alpha[z]$. We need to show $\hat{C}^0[y]$. Suppose $z = 0$. Then $y = Sx$ by (3.3). By (3) we have $\hat{C}^0[x]$ and from (2) we obtain $\hat{C}^0[Sx]$, i.e. $\hat{C}^0[y]$. Thus $z = 0 \to \hat{C}^0[y]$, so suppose $z \neq 0$. Then $SPz = z$ by (3.7), and so $S(y + Pz) = y + SPz = y + z = Sx$ by (3.4). Consequently $y + Pz = x$, by (3.2). But since $\alpha[z]$ we have $\alpha[Pz]$ by (6.3), and since $\hat{C}^1[x]$ we have $\hat{C}^0[y]$. Thus $z \neq 0 \to \hat{C}^0[y]$, and so $\hat{C}^0[y]$. Thus (5).

6. $\hat{C}^1[x]$ & $u+v = x$ & $\alpha[v] \to \hat{C}^1[u]$. Suppose *hyp* (6) & $y+z = u$ & $\alpha[z]$. We need to show $\hat{C}^0[y]$. We have $(y + z) + v = x$, and since $\alpha[v]$ we obtain $y + (z + v) = x$. But $\alpha[z + v]$ by (6.4), and since $\hat{C}^1[x]$ we get $\hat{C}^0[y]$. Thus *hyp* (6) $\to \hat{C}^1[u]$, i.e. (6).

7. $\hat{C}^2[x] \to \hat{C}^1[x]$. Suppose $\hat{C}^2[x]$. We have $\hat{C}^1[0]$ by (4), so we have $\hat{C}^1[0+x]$. Therefore $\hat{C}^0[0+x]$ by (3), and hence $\delta[0+x]$. By (6.12), $x = 0+x$ and so $\hat{C}^1[x]$. Thus (7).

8. $\hat{C}^2[0]$, by (3.3).

9. $\hat{C}^2[x] \to \hat{C}^2[Sx]$. Suppose $\hat{C}^2[x]$ & $\hat{C}^1[y]$. We need to show $\hat{C}^1[y + Sx]$. But $\hat{C}^1[y + x]$, and so $\hat{C}^1[S(y + x)]$ by (5). By (3.4) we find $\hat{C}^1[y + Sx]$, and thus (9).

10. $\hat{C}^2[x]$ & $u + v = x$ & $\alpha[v] \to \hat{C}^2[u]$. Suppose *hyp* (10) & $\hat{C}^1[y]$. We need to show $\hat{C}^1[y + u]$. But $(y + u) + v = y + (u + v)$ since $\alpha[v]$, and so $(y + u) + v = y + x$. We have $\hat{C}^1[y + x]$, since $\hat{C}^2[x]$ & $\hat{C}^1[y]$, and we have $\alpha[v]$. Hence $\hat{C}^1[y + u]$ by (6), and thus (10).

11. $\hat{C}^2[x_1]$ & $\hat{C}^2[x_2] \to \hat{C}^2[x_1 + x_2]$. Suppose $\hat{C}^2[x_1]$ & $\hat{C}^2[x_2]$ & $\hat{C}^1[y]$. We need to show $\hat{C}^1[y + (x_1 + x_2)]$. But by (7) and (3) we have $\hat{C}^0[x_2]$ and therefore $\alpha[x_2]$, so that $y + (x_1 + x_2) = (y + x_1) + x_2$. We get $\hat{C}^1[y + x_1]$ from $\hat{C}^2[x_1]$, and therefore we get $\hat{C}^1[(y + x_1) + x_2]$ from $\hat{C}^2[x_2]$. That is, $\hat{C}^1[y + (x_1 + x_2)]$. Thus (11).

12. $\hat{C}^3[x] \to \hat{C}^2[x]$, by definition.

13. $\hat{C}^3[0]$, by (8) and (3.5).

14. $\hat{C}^3[x] \to \hat{C}^3[Sx]$. Suppose $\hat{C}^3[x]$. Then $\hat{C}^2[Sx]$ by (12) and (9). Suppose $\hat{C}^2[y]$. We need to show $\hat{C}^2[y \cdot Sx]$. But $y \cdot Sx = y \cdot x + y$ by (3.6), and we have $\hat{C}^2[y \cdot x]$, so by (11) we get $\hat{C}^2[y \cdot x + y]$, i.e. $\hat{C}^2[y \cdot Sx]$. Thus

$\hat{C}^2[y] \to \hat{C}^2[y \cdot Sx]$, and thus (14).

15. $\hat{C}^3[x] \to \hat{C}^3[Px]$. We have $x = 0 \to (15)$ by (3.7), so suppose $x \neq 0$ & $\hat{C}^3[x]$. First we need to show $\hat{C}^2[Px]$. But $Px + S0 = S(Px + 0) = SPx = x$ by (3.4), (3.3), and (3.7), and $\alpha[S0]$ by (6.1) and (6.2). By (12) we have $\hat{C}^2[x]$. By (10) applied to $Px + S0 = x$ we have $\hat{C}^2[Px]$. Now suppose $\hat{C}^2[y]$. We need to show $\hat{C}^2[y \cdot Px]$. But $y \cdot Px + y = y \cdot SPx = y \cdot x$ by (3.6) and (3.7). Since $\hat{C}^2[y]$, we have $\hat{C}^0[y]$ by (7) and (3), and so $\alpha[y]$. Since $\hat{C}^3[x]$ & $\hat{C}^2[y]$, we have $\hat{C}^2[y \cdot x]$. By (10) applied to $y \cdot Px + y = y \cdot x$ we have $\hat{C}^2[y \cdot Px]$. Thus $\hat{C}^2[y] \to \hat{C}^2[y \cdot Px]$, and thus $x \neq 0 \to (15)$. Consequently we have (15).

16. $\hat{C}^3[x_1]$ & $\hat{C}^3[x_2] \to \hat{C}^3[x_1 + x_2]$. Suppose *hyp*(16). Then we have $\hat{C}^2[x_1]$ & $\hat{C}^2[x_2]$ by (12), so $\hat{C}^2[x_1 + x_2]$ by (11). Suppose $\hat{C}^2[y]$. We need to show $\hat{C}^2[y \cdot (x_1 + x_2)]$. By (7) and (3) we have $\hat{C}^0[y]$ and $\hat{C}^0[x_2]$, so that $\alpha[y]$ and $\beta[x_2]$. Hence $y \cdot (x_1 + x_2) = y \cdot x_1 + y \cdot x_2$. We have $\hat{C}^2[y \cdot x_1]$ and $\hat{C}^2[y \cdot x_2]$, so by (11) we have $\hat{C}^2[y \cdot x_1 + y \cdot x_2]$, i.e. $\hat{C}^2[y \cdot (x_1 + x_2)]$. Thus $\hat{C}^3[x_1 + x_2]$, and thus (16).

17. $\hat{C}^3[x_1]$ & $\hat{C}^3[x_2] \to \hat{C}^3[x_1 \cdot x_2]$. Suppose *hyp*(17). By definition of $\hat{C}^3$, we have $\hat{C}^2[x_1]$ and, since $\hat{C}^3[x_2]$, we have $\hat{C}^2[x_1 \cdot x_2]$. Now suppose $\hat{C}^2[y]$; we want to show $\hat{C}^2[y \cdot (x_1 \cdot x_2)]$. We have $\hat{C}^0[x_2]$, $\hat{C}^0[x_1]$, and $\hat{C}^0[y]$ by (12), (7), and (3). Therefore $\gamma[x_2]$, $\beta[x_1]$, and $\alpha[y]$, so that $y \cdot (x_1 \cdot x_2) = (y \cdot x_1) \cdot x_2$. From $\hat{C}^3[x_1]$ we get $\hat{C}^2[y \cdot x_1]$, and then from $\hat{C}^3[x_2]$ we get $\hat{C}^2[(y \cdot x_1) \cdot x_2]$, i.e. $\hat{C}^2[y \cdot (x_1 \cdot x_2)]$. Thus $\hat{C}^2[y] \to \hat{C}^2[y \cdot (x_1 \cdot x_2)]$, and thus (17).

We have $\hat{C}^3[x] \to C[x]$ by (12), (7), and (3), so by (13), (14), (15), (16), and (17) we have established *con*($\widehat{\text{REL}}$). □

Whenever we encounter an open formula of Q_0 that is inductive in one of its variables, we can use $\widehat{\text{REL}}$ to refine our concept of number and adjoin the formula as a new axiom, obtaining an open theory that is interpretable in Q_0, and the process can be iterated. This observation is expressed in greater detail in the following two metatheorems.

Metatheorem 6.2 Let T *be an open extension of* Q_0 *with the same language, and let* A *be an open formula that is inductive in* x *in* T. *Then* T[A] *is interpretable in* T.

Demonstration. Let C be $A_{/x}$. Then $\hat{C}^3$ respects each function symbol (0, S, P, +, and ·) of T, by $\widehat{\text{REL}}$. It respects all of the axioms of T because they are open, and it respects A because $\hat{C}^3$ is stronger than $A_{/x}$, by $\widehat{\text{REL}}$.

Consequently, T[A] is interpretable in T via the interpretation associated with $\hat{C}^3$. □

Let $\tilde{Q}_0$ be the formal system with the same language, formulas, and axioms as Q_0, but with an additional rule of inference (open induction):

OI. If A is open and x occurs in A, infer A from ind_x A.

Metatheorem 6.3 Let $B_1, \ldots, B_\lambda$ *be theorems of* $\tilde{Q}_0$. *Then the theory* $Q_0[B_1, \ldots, B_\lambda]$ *is interpretable in* Q_0.

Demonstration. Let B be B_1 & $\cdots$ & B_λ, so that B is a theorem of $\tilde{Q}_0$. Let $A_1, \ldots, A_\nu$ be the open formulas occurring in the proof in $\tilde{Q}_0$ of B that are inferred by OI, in the order of their occurrence, and let $x_1, \ldots, x_\nu$ be the corresponding variables. Let T_1 be $Q_0[A_1]$, let T_2 be $Q_0[A_1, A_2]$, ..., and let T_ν be $Q_0[A_1, \ldots, A_\nu]$. Then

$$\vdash_{Q_0} ind_{x_1} A_1, \ \vdash_{T_1} ind_{x_2} A_2, \ \ldots, \ \vdash_{T_{\nu-1}} ind_{x_\nu} A_\nu.$$

By Metatheorem 6.2 applied ν times, T_1 is interpretable in Q_0, T_2 is interpretable in T_1, ..., and T_ν is interpretable in $T_{\nu-1}$. By the Interpretation Theorem of [Sh,§4.7], T_ν is interpretable in Q_0 and, since $B_1, \ldots, B_\lambda$ are theorems of T_ν, the theory $Q_0[B_1, \ldots, B_\lambda]$ is interpretable in Q_0.

Metatheorem 6.4 Q_1' *is interpretable in* Q.

Demonstration. Q_1' is an extension by definitions of Q_1, and Q_0 is equivalent to an extension by definitions of Q, so it suffices to show that Q_1 is interpretable in Q_0. To do this it is enough, by Metatheorem 6.3, to prove (3.8)–(3.12) in $\tilde{Q}_0$. We do this now.

We have (3.8) by (6.1), (6.2), and OI. We have (3.9) by (3.8), (6.5), (6.6), and OI. We have (3.10) by (3.8), (3.9), (6.7), (6.8), and OI.

We claim that 1: $0 + x = x$. We have $0 + 0 = 0$ by (3.3). Suppose $0 + x = x$. Then $0 + \mathrm{S}x = \mathrm{S}(0 + x) = \mathrm{S}x$ by (3.4), and thus $0 + x = x \rightarrow 0 + \mathrm{S}x = \mathrm{S}x$. By OI we have (1). We claim that 2: $\mathrm{S}y + x = \mathrm{S}(y + x)$. We have $\mathrm{S}y + 0 = \mathrm{S}y = \mathrm{S}(y + 0)$ by (3.3). Suppose $\mathrm{S}y + x = \mathrm{S}(y + x)$. Then $\mathrm{S}x + \mathrm{S}y = \mathrm{S}(\mathrm{S}y + x) = \mathrm{SS}(y + x) = \mathrm{S}(y + \mathrm{S}x)$ by (3.4), and thus $(2) \rightarrow (2)_x[\mathrm{S}x]$. By OI we have (2). We claim that (3.11), namely $x + y = y + x$. We have $x + 0 = 0 + x$ by (3.3) and (1). Suppose $x + y = y + x$. By (2) and (3.4) we obtain $\mathrm{S}y + x = \mathrm{S}(y + x) = \mathrm{S}(x + y) = x + \mathrm{S}y$, and thus $(3.11) \rightarrow (3.11)_y[\mathrm{S}y]$. By OI we have (3.11).

We claim that 3: $0 \cdot x = 0$. We have $0 \cdot 0 = 0$ by (3.5). Suppose $0 \cdot x = 0$. Then $0 \cdot \mathrm{S}x = 0 \cdot x + 0 = 0 \cdot x = 0$ by (3.6) and (3.3), and thus $(3) \to (3)_x[\mathrm{S}x]$. By OI we have (3). We claim that 4: $\mathrm{S}x \cdot y = x \cdot y + y$. We have $\mathrm{S}x \cdot 0 = 0$ by (3.5) and $x \cdot 0 + 0 = x \cdot 0 = 0$ by (3.3) and (3.5), so that $\mathrm{S}x \cdot 0 = x \cdot 0 + 0$. Suppose $\mathrm{S}x \cdot y = x \cdot y + y$. Then by (3.6), (3.8), (3.4), and (3.11) we obtain $\mathrm{S}x \cdot \mathrm{S}y = \mathrm{S}x \cdot y + \mathrm{S}x = (x \cdot y + y) + \mathrm{S}x = x \cdot y + (y + \mathrm{S}x) = x \cdot y + \mathrm{S}(y + x) = x \cdot y + \mathrm{S}(x + y) = x \cdot y + (x + \mathrm{S}y) = (x \cdot y + x) + \mathrm{S}y = x \cdot \mathrm{S}y + \mathrm{S}y$. Thus $(4) \to (4)_y[\mathrm{S}y]$. By OI we have (4). Finally, we claim that (3.12), namely $x \cdot y = y \cdot x$. We have $0 \cdot y = y \cdot 0$ by (3) and (3.5). Suppose $x \cdot y = y \cdot x$. By (4) and (3.6), $\mathrm{S}x \cdot y = x \cdot y + y = y \cdot x + y = y \cdot \mathrm{S}x$. Thus $(3.12) \to (3.12)_x[\mathrm{S}x]$. By OI we have (3.12).

Chapter 7

Bounded induction

In this chapter we define the notion of bounded formulas and show that induction can be used on them.

The initial occurrence of $\exists$x in a part $\exists$xB of A is called *manifestly bounded* in case B is of the form x $\leq$ a & C where a is a term not containing x. The formula A is called *manifestly bounded* in case each occurrence of an existential quantifier in A is manifestly bounded. (Recall that inside each universal quantifier there lurks an existential quantifier.) Formulas are built up from atomic formulas by means of $\neg$, $\vee$, and existential quantifiers $\exists$x (see [Sh,§2.4]). In the same way, manifestly bounded formulas are built up from atomic formulas by means of $\neg$, $\vee$, and manifestly bounded occurrences of existential quantifiers $\exists$x.

Metatheorem 7.1 Let T *be a theory containing the binary predicate symbol* $\leq$*, and let* C *be a unary formula of an extension* U *of* T *that respects all function symbols of* T *and is hereditary. Let* A *be a manifestly bounded formula of* T*. Then*

$$\vdash_U C(free\,A) \rightarrow (A \leftrightarrow A_C).$$

Demonstration. For the purposes of this demonstration, call a formula B *good* in case it is a formula of T and $\vdash_U C(free\,B) \rightarrow (B \leftrightarrow B_C)$. Atomic formulas of T are good. Suppose B is good. Since $C(free\,\neg B)$ is $C(free\,B)$ and $(\neg B)_C$ is $\neg(B_C)$, the formula $\neg$B is good. Thus 1: if B is good then $\neg$B is good. Suppose B_1 and B_2 are good. Since

$$\mathrm{C}(\mathit{free}\,(\mathrm{B}_1 \vee \mathrm{B}_2)) \rightarrow \mathrm{C}(\mathit{free}\,\mathrm{B}_1)\ \&\ \mathrm{C}(\mathit{free}\,\mathrm{B}_2)$$

is a tautology or the empty expression, and since $(\mathrm{B}_1 \vee \mathrm{B}_2)_\mathrm{C}$ is $\mathrm{B}_{1\mathrm{C}} \vee \mathrm{B}_{2\mathrm{C}}$, we have that $\mathrm{B}_1 \vee \mathrm{B}_2$ is good. Thus 2: if B_1 and B_2 are good then $\mathrm{B}_1 \vee \mathrm{B}_2$ is good. Suppose D is good and B is $\exists \mathrm{x}(\mathrm{x} \leq \mathrm{a}\ \&\ \mathrm{D})$ and a does not contain x. Then every variable in a occurs free in B, and since C respects every function symbol of T we have $\vdash_\mathrm{U} \mathrm{C}(\mathit{free}\,\mathrm{B}) \rightarrow \mathrm{C}[\mathrm{a}]$. Since C is hereditary,

$$\vdash_\mathrm{U} \mathrm{C}(\mathit{free}\,\mathrm{B}) \rightarrow (\exists \mathrm{x}(\mathrm{x} \leq \mathrm{a}\ \&\ \mathrm{D}) \leftrightarrow \exists \mathrm{x}(\mathrm{C}[\mathrm{x}]\ \&\ \mathrm{x} \leq \mathrm{a}\ \&\ \mathrm{D})).$$

Since D is good and $\mathrm{C}(\mathit{free}\,\mathrm{B})\ \&\ \mathrm{C}[\mathrm{x}] \rightarrow \mathrm{C}(\mathit{free}\,\mathrm{D})$ is a tautology (unless D is closed, in which case the question does not arise),

$$\vdash_\mathrm{U} \mathrm{C}(\mathit{free}\,\mathrm{B}) \rightarrow (\exists \mathrm{x}(\mathrm{x} \leq \mathrm{a}\ \&\ \mathrm{D}) \leftrightarrow \exists \mathrm{x}(\mathrm{C}[\mathrm{x}]\ \&\ \mathrm{x} \leq \mathrm{a}\ \&\ \mathrm{D}_\mathrm{C})),$$

which is $\vdash_\mathrm{U} \mathrm{C}(\mathit{free}\,\mathrm{B}) \rightarrow (\mathrm{B} \leftrightarrow \mathrm{B}_\mathrm{C})$. Thus 3: if D is good and a does not contain x then $\exists \mathrm{x}(\mathrm{x} \leq \mathrm{a}\ \&\ \mathrm{D})$ is good. Starting from the atomic formulas in A and applying (1), (2), and (3) to each occurrence of $\neg$, $\vee$, or $\exists$ in A, we find that A is good.

Metatheorem 7.2 Let A *be a manifestly bounded formula of* Q_1' *that is inductive in one of its free variables* x. *Then* $Q_1'[\mathrm{A}]$ *is interpretable in* Q_1'.

Demonstration. Let C be $\mathrm{A}_{/\mathrm{x}}$. By REL, C^3 respects all function symbols and nonlogical axioms of Q_1'. Since C^3 is stronger than $\mathrm{A}_{/\mathrm{x}}$ by REL, we have $\vdash_{Q_1'} \mathrm{C}^3(\mathit{free}\,\mathrm{A}) \rightarrow \mathrm{A}$. Since C^3 is hereditary by REL, C^3 respects A by Metatheorem 7.1. Therefore $Q_1'[\mathrm{A}]$ is interpretable in Q_1' via the interpretation associated with C^3. □

We construct a theory Q_2, which is an extension of Q_1' with the same language, by adjoining as new nonlogical axioms all formulas of the form

$$\text{MBI. } \mathrm{A}_x[0]\ \&\ \neg\exists x(x \leq y\ \&\ \neg\mathrm{A}_x[\mathrm{S}x]) \rightarrow (x \leq y \rightarrow \mathrm{A})$$

where A is a manifestly bounded formula in the language of Q_1' and y does not occur free in A.

Metatheorem 7.3 Let $\mathrm{B}_1, \ldots, \mathrm{B}_\lambda$ *be theorems of* Q_2. *Then* $Q_1'[\mathrm{B}_1, \ldots, \mathrm{B}_\lambda]$ *is interpretable in* Q_1'.

Demonstration. Each of the new nonlogical axioms (MBI) is manifestly bounded. We claim that each (MBI) is inductive in y in Q_1'. To see this, we argue in Q_1' as follows.

Since $x \leq 0 \rightarrow x = 0$ by (4.3), we have $\mathrm{A}_x[0] \rightarrow (x \leq 0 \rightarrow \mathrm{A})$, and therefore $(\mathrm{MBI})_y[0]$. Suppose (MBI) & *hyp* $(\mathrm{MBI})_y[\mathrm{S}y]$. Observe that *hyp* $(\mathrm{MBI})_y[\mathrm{S}y] \rightarrow$ *hyp* (MBI) since, by (4.4), $x \leq y \rightarrow x \leq \mathrm{S}y$. Therefore we have *hyp* (MBI), and since we have (MBI) we also have *con* (MBI), namely 1: $x \leq y \rightarrow \mathrm{A}$. We need to show that 2: $x \leq \mathrm{S}y \rightarrow \mathrm{A}$. But by (4.4), $x \leq \mathrm{S}y \rightarrow x \leq y \vee x = \mathrm{S}y$. By (1) we have $x \leq y \rightarrow$ (2), so suppose $x = \mathrm{S}y$. Since *hyp* (MBI), we have—rewriting it in a more civilized notation—

$$\mathrm{A}_x[0] \ \& \ \forall x(x \leq y \ \& \ \mathrm{A} \rightarrow \mathrm{A}_x[\mathrm{S}x]).$$

But $y \leq y$ by (4.2), so $\mathrm{A}_x[y] \rightarrow \mathrm{A}_x[\mathrm{S}y]$. Also because $y \leq y$, we have $\mathrm{A}_x[y]$ by (1). Hence $\mathrm{A}_x[\mathrm{S}y]$, i.e. A, and thus (2). Thus $(\mathrm{MBI}) \rightarrow (\mathrm{MBI})_y[\mathrm{S}y]$, and therefore ind_y (MBI).

Since each (MBI) is manifestly bounded and inductive in y in Q_1', the conjunction B of all axioms of the form (MBI) occurring in the proofs in Q_2 of $\mathrm{B}_1, \ldots, \mathrm{B}_\lambda$ is also manifestly bounded and inductive in y in Q_1'. By Metatheorem 7.2, $Q_1'[\mathrm{B}]$ is interpretable in Q_1'. But $\mathrm{B}_1, \ldots, \mathrm{B}_\lambda$ are theorems of $Q_1'[\mathrm{B}]$, so by the Interpretation Theorem of [Sh,§4.7], $Q_1'[\mathrm{B}_1, \ldots, \mathrm{B}_\lambda]$ is interpretable in Q_1'. □

We will say that a theory T′ is *locally interpretable* in a theory T in case whenever $\mathrm{B}_1, \ldots, \mathrm{B}_\lambda$ are theorems of T′, then the theory whose nonlogical axioms are $\mathrm{B}_1, \ldots, \mathrm{B}_\lambda$ (and whose nonlogical symbols are those occurring in $\mathrm{B}_1, \ldots, \mathrm{B}_\lambda$) is interpretable in T. By Metatheorems 7.3 and 6.4, the theory Q_2 is locally interpretable in Q. I promised in Chapter 3 that we would work only in theories interpretable in Q; since only finitely many axioms of Q_2 will ever be used in this investigation, that promise will be kept so long as we work in Q_2 or extensions by definition of Q_2. Moreover, Pudlák sketches a proof of the result that Q_2 is globally interpretable in Q (Theorem 2.7 of [Pu], attributed to A. Wilkie). It would be awkward always to write formulas in manifestly bounded form. For example, neither $\forall x(x \leq y \rightarrow x \leq z)$ nor $\exists x \exists y(x \leq z \ \& \ y \leq z \ \& \ x \neq y)$ is manifestly bounded, though they are respectively equivalent to the manifestly bounded formulas $\neg\exists x(x \leq y \ \& \ \neg x \leq z)$ and $\exists x(x \leq z \ \& \ \exists y(y \leq z \ \& \ x \neq y))$. We use

$$\mathrm{A}\colon \mathrm{x}_1 \leq \mathrm{a}_1, \ldots, \mathrm{x}_\nu \leq \mathrm{a}_\nu$$

as an abbreviation for $\mathrm{A} \leftrightarrow \mathrm{A}'$, where A′ is the formula obtained by replacing each part of A of the form $\exists \mathrm{x}_\mu \mathrm{B}$ by $\exists \mathrm{x}_\mu(\mathrm{x}_\mu \leq \mathrm{a}_\mu \ \& \ \mathrm{B})$, for all μ from

1 to ν. Let T be a theory containing $\leq$, let A be a formula of T, and let $x_1, \ldots, x_\nu$ be the variables that occur bound in A. We say that A is of *bounded form* (or *of bounded form in* T) in case there are terms $a_1, \ldots, a_\nu$ such that for all μ from 1 to ν, the variable x_μ does not occur in a_μ, and $\vdash_T A\colon x_1 \leq a_1, \ldots, x_\nu \leq a_\nu$. Then the formula A′ defined as above is manifestly bounded, and A is equivalent to A′ in T. An extension T′ of T is called a *bounded extension* of T in case it is an extension by definitions of T such that for each defining axiom $px_1 \ldots x_\nu \leftrightarrow D$ of a predicate symbol, D is of bounded form, and for each defining axiom $fx_1 \ldots x_\nu = y \leftrightarrow D$ of a function symbol, $\exists y D$ is of bounded form. A nonlogical symbol, term, or formula of an extension U of T is called *bounded over* T in case it is a nonlogical symbol, term, or formula of bounded form of a bounded extension T′ of T (such that U is an extension of T′).

Let T be a theory containing $\leq$, 0, and S. We say that T is a *bounded theory* in case for every formula A of bounded form in T, $\vdash_T \mathit{ind}_x A \rightarrow A$.

Metatheorem 7.4 Q_2 is a bounded theory.

Demonstration. Let A be a formula of bounded form in Q_2, so that it is equivalent to the manifestly bounded formula A′. Then we argue in Q_2 as follows.

Suppose $\mathit{ind}_x A'$; that is, $A'_x[0]\ \&\ \forall x(A' \rightarrow A'_x[Sx])$. Then

$$A'_x[0]\ \&\ \neg\exists x(x \leq y\ \&\ A'\ \&\ \neg A'_x[Sx]).$$

Here y is a variable distinct from x and all variables occurring in A′. By (MBI) we have $x \leq y \rightarrow A'$. Therefore $x \leq x \rightarrow A'$, and so A′ by (4.2). Thus $\mathit{ind}_x A' \rightarrow A'$, and so $\mathit{ind}_x A \rightarrow A$.

Metatheorem 7.5 Let T *be a bounded theory and let* T′ *be a bounded extension of* T. *Then* T′ *is a bounded theory.*

Demonstration. Suppose first that T′ is obtained by adjoining a single symbol, and let A be a formula of bounded form in T′. Then the translation A* of A into T (see [Sh,§4.6]) is a formula of bounded form in T, so that $\vdash_T \mathit{ind}_x A^* \rightarrow A^*$. But $\vdash_{T'} A^* \leftrightarrow A$, and so $\vdash_{T'} \mathit{ind}_x A \rightarrow A$. Thus the result holds in this case. To obtain the result in general, apply this result step by step to each new nonlogical symbol. □

It might appear simpler to argue in the general case directly, by considering the translation A* into T. But our metamathematical arguments are intended to apply to actual formulas that one writes down. The metatheorems are intended to be correct statements about what one can actually do, and the demonstrations are intended to show how to actually carry out the constructions they assert to exist. With even a rather modest extension by definitions T′ of T, if we try to construct the translation A^* of a formula A of T′ all the way into T, we may find ourselves in the plight of the legendary caliph who promised to give one grain of wheat for the first square of a chessboard and to double the number for each successive square.

Metatheorem 7.6 Let U *be an extension of* Q_2 *and let* A *be a formula of* U *that is bounded over* Q_2. *Then*

$$\text{BI.} \quad \textit{ind}_x\, A \to A$$

is a theorem of U.

Demonstration. By Metatheorems 7.4 and 7.5. □

We refer to BI as *bounded induction.*

Metatheorem 7.7 Let T *be a theory containing* $\leq$ *and* 0, *let* T′ *be a bounded extension of* T, *and let* U *be an extension of* T′. *Let* C *be a unary formula of* U *that is hereditary and respects each nonlogical axiom and function symbol of* T. *Then* C *respects each nonlogical axiom and function symbol of* T′.

Demonstration. Suppose first that T′ is obtained from T by adjoining the new predicate symbol p with defining axiom $px_1 \ldots x_\nu \leftrightarrow D$, where D is a formula of bounded form in T. Then D is equivalent to the manifestly bounded formula D′ of T; we may assume that D′ contains no free variables other than $x_1, \ldots, x_\nu$, for if it did, we could substitute 0 for them. By Metatheorem 7.1, C respects $px_1 \ldots x_\nu \leftrightarrow D'$, and so C respects $px_1 \ldots x_\nu \leftrightarrow D$. Thus the result holds in this case.

Suppose next that T′ is obtained from T by adjoining the new function symbol f with defining axiom $fx_1 \ldots x_\nu = y \leftrightarrow D$, where $\exists yD$ is a formula of bounded form in T. Then D is also a formula of bounded form in T, and so is equivalent to the manifestly bounded formula D′ of T; we may assume that D′ contains no free variables other than $x_1, \ldots, x_\nu, y$. By Metatheorem 7.1, C respects $fx_1 \ldots x_\nu = y \leftrightarrow D'$, and so C respects $fx_1 \ldots x_\nu = y \leftrightarrow D$. Now $\exists yD$ is equivalent to a formula $\exists y(y \leq a \;\&\; D_0)$, where a is a term of T not containing y; we may assume that a contains no variables other than $x_1, \ldots, x_\nu$. By the uniqueness condition,

$$fx_1 \ldots x_\nu = y \leftrightarrow y \leq a \mathbin{\&} D_0.$$

But since C respects each function symbol of T, we have

$$\vdash_U C[x_1] \to \cdots \to C[x_\nu] \to C[a],$$

and since C is hereditary it respects f. Thus the result holds in this case.

To obtain the result in general, apply these results step by step to each new nonlogical symbol.

Chapter 8

The bounded least number principle

The least number principle is a form of the induction principle that is useful in many proofs. Used indirectly, it is equivalent to proof by infinite descent or complete induction. Here we formulate a predicative version of the least number principle.

For a formula A and variables $x_1, \ldots, x_\nu$, let $y_1, \ldots, y_\nu$ be in alphabetical order the first ν variables not occurring in A and distinct from $x_1, \ldots, x_\nu$; then we write

$$\min_{x_1 \ldots x_\nu} A$$

for

$$A \mathbin{\&} \neg \exists y_1 \ldots \exists y_\nu (y_1 \leq x_1 \mathbin{\&} \cdots \mathbin{\&} y_\nu \leq x_\nu \mathbin{\&} \\ (y_1 \neq x_1 \vee \cdots \vee y_\nu \neq x_\nu) \mathbin{\&} A_{x_1 \ldots x_\nu}[y_1 \ldots y_\nu]).$$

Metatheorem 8.1 Let U *be an extension of* Q_2 *and let* A *be a formula of* U *that is bounded over* Q_2. *Then*

$$\text{BLNP. } \exists x_1 \cdots \exists x_\nu\, A \rightarrow \exists x_1 \cdots \exists x_\nu \min_{x_1 \ldots x_\nu} A$$

is a theorem of U.

Demonstration. Consider first the case that ν is 1. We use the abbreviations

α for $\exists x_1(x_1 \leq z \;\&\; A)$,

β for $\exists x_1(x_1 \leq z \;\&\; \min_{x_1} A)$,

where z is a variable distinct from x_1 and y_1 that does not occur in A. Then we argue in U as follows. We claim that 1: $\alpha \rightarrow \beta$. Suppose $\alpha_z[0]$. Then there exists x_1 such that $x_1 \leq 0 \;\&\; A$. By (4.3), $x_1 = 0$ and so $\min_{x_1} A$. Thus $(1)_z[0]$. By (4.2) and (4.5) we have 2: $\beta \rightarrow \beta_z[Sz]$. Suppose $\alpha_z[Sz]$, so that there exists x_1 such that $x_1 \leq Sz \;\&\; A$. By (4.4), $x_1 \leq z \vee x_1 = Sz$. We have $x_1 \leq z \rightarrow \alpha$. Suppose $\neg\alpha$. Then $x_1 = Sz$. Suppose $\neg(\min_{x_1} A)_{x_1}[Sz]$. Then there exists y_1 such that $y_1 \leq Sz \;\&\; y_1 \neq Sz \;\&\; A_{x_1}[y_1]$. By (4.4), $y_1 \leq z$, so that α, a contradiction. Thus $(\min_{x_1} A)_{x_1}[Sz]$. By (4.2), $Sz \leq Sz$, and thus $\neg\alpha \rightarrow \beta_z[Sz]$. Thus 3: $\alpha_z[Sz] \rightarrow \alpha \vee \beta_z[Sz]$. By (2) and (3),

$$(\alpha \rightarrow \beta) \;\&\; \alpha_z[Sz] \rightarrow \beta_z[Sz],$$

so that $(1) \rightarrow (1)_z[Sz]$. By BI we have (1). Consequently $\alpha_z[x_1] \rightarrow \beta_z[x_1]$, but since $x_1 \leq x_1$ by (4.2), we have $\exists x_1 A \rightarrow \exists x_1 \min_{x_1} A$. This proves the result when ν is 1.

Now consider the general case. Let $z_1, \ldots, z_\nu$ be distinct, be distinct from $x_1, \ldots, x_\nu, y_1, \ldots, y_\nu$, and not occur in A. We argue in U as follows. Suppose *hyp*(BLNP). Then there exist $z_1, \ldots, z_\nu$ such that $A_{x_1 \ldots x_\nu}[z_1 \ldots z_\nu]$. Write

$$\gamma \text{ for } x_1 \leq z_1 \;\&\; \cdots \;\&\; x_\nu \leq z_\nu \;\&\; A.$$

By (4.2), $\gamma_{x_1 \ldots x_\nu}[z_1 \ldots z_\nu]$. By the result when ν is 1, applied ν times,

there exists x_1 such that $\min_{x_1} \exists x_2 \cdots \exists x_\nu\, \gamma$,

there exists x_2 such that $\min_{x_2} \exists x_3 \cdots \exists x_\nu\, \gamma$,

...

there exists x_ν such that $\min_{x_\nu} \gamma$.

Suppose $\neg\min_{x_1 \ldots x_\nu} A$. Since γ, we have A, so there exist $y_1, \ldots, y_\nu$ such that

$y_1 \leq x_1 \;\&\; \cdots \;\&\; y_\nu \leq x_\nu \;\&\; (y_1 \neq x_1 \vee \cdots \vee y_\nu \neq x_\nu) \;\&$
$A_{x_1,\ldots,x_\nu}[y_1, \ldots, y_\nu]$.

For some μ with $1 \leq \mu \leq \nu$ we have $y_\mu \neq x_\mu$, but this contradicts $\min_{x_\mu} \exists x_{\mu+1} \cdots \exists x_\nu \gamma$. Thus $\min_{x_1 \ldots x_\nu} A$, and thus (BLNP). □

We will write "$\exists x_1 \cdots \exists x_\nu A$. By BLNP there exist minimal such $x_1, \ldots,$ and x_ν" for "$\exists x_1 \cdots \exists x_\nu A$. By BLNP there exist $x_1, \ldots,$ and x_ν such that $\min_{x_1 \ldots x_\nu} A$" (with the appropriate change in grammar or punctuation when ν is 1 or 2).

Chapter 9

The Euclidean algorithm

Until further notice we will work in bounded extensions of Q_2. At one point in the proof of REL we quoted five axioms to show that one times x equals x. This sort of thing might become tiresome if continued much longer, so let's stop doing it. The development picks up from where we left off at the end of Chapter 4, but now we have BI and BLNP available.

9.1 *Thm.* $z_1 + y = z_2 + y \rightarrow z_1 = z_2$.

Proof. Clearly $(9.1)_y[0]$. Suppose (9.1) & $z_1 + Sy = z_2 + Sy$. Then $z_1 + y = z_2 + y$ and so $z_1 = z_2$. Thus $ind_y\,(9.1)$, so by BI we have (9.1).

9.2 *Def.* $x - y = z \leftrightarrow z + y = x$, otherwise $z = 0$.

The uniqueness condition holds by (9.1). We have $\exists z\; rhs\,(9.2)\colon z \leq x$, so (9.2) is the defining axiom of a bounded function symbol. Put less formally, this function symbol is bounded because we do not need to search through all numbers to see whether there is a z with $z + y = x$, but only through all numbers $\leq x$.

9.3 *Thm.* $y \neq 0 \;\&\; z \cdot y = 0 \rightarrow z = 0$.

Proof. Suppose $y \neq 0 \;\&\; z \cdot y = 0 \;\&\; z \neq 0$. Then $z = \mathrm{SP}z$, so $\mathrm{SP}z \cdot y = 0$, $\mathrm{P}z \cdot y + y = 0$, and $y = 0$, which is a contradiction. Thus (9.3).

9.4 *Thm.* $y \neq 0 \;\&\; z_1 \cdot y = z_2 \cdot y \rightarrow z_1 = z_2$.

Proof. Suppose $\exists z_1 \exists z_2 \neg(9.4)$. By BLNP there exist minimal such z_1 and z_2. Of course, $\neg(9.4) \rightarrow hyp\,(9.4)$, so we have $hyp\,(9.4)$. By (9.3),

$z_1 \neq 0$ and $z_2 \neq 0$. Hence $\mathrm{SP}z_1 \cdot y = \mathrm{SP}z_2 \cdot y$, $\mathrm{P}z_1 \cdot y + y = \mathrm{P}z_2 \cdot y + y$, and $\mathrm{P}z_1 y = \mathrm{P}z_2 \cdot y$ by (9.1). By the minimality assumption, $\mathrm{P}z_1 = \mathrm{P}z_2$. Therefore $\mathrm{SP}z_1 = \mathrm{SP}z_2$ and we have $z_1 = z_2$, a contradiction. Thus (9.4).

9.5 *Thm.* $y \neq 0 \rightarrow z \leq z \cdot y$.

Proof. Suppose $y \neq 0$. Then $z \cdot y = z \cdot \mathrm{SP}y = z \cdot \mathrm{P}y + z$, so $z \leq z \cdot y$. Thus (9.5).

9.6 *Def.* $x/y = z \leftrightarrow z \cdot y = x \;\&\; y \neq 0$, otherwise $z = 0$.

The uniqueness condition holds by (9.4), and we have $\exists z\, rhs\,(9.6)$: $z \leq x$ by (9.5).

9.7 *Thm.* $x \leq y \vee y \leq x$.

Proof. We have $(9.7)_x[0]$ since $0 \leq y$ by (4.2). Suppose $y \leq x$. Then there exists z such that $y + z = x$, so $y + \mathrm{S}z = \mathrm{S}x$. Thus $y \leq x \rightarrow y \leq \mathrm{S}x$. By (4.2), $y = x \rightarrow y \leq \mathrm{S}x$, so suppose $x \leq y \;\&\; x \neq y$. Then there exists z such that $x + z = y \;\&\; z \neq 0$, so that $x + \mathrm{SP}z = y$, $\mathrm{S}x + \mathrm{P}z = y$, and $\mathrm{S}x \leq y$. Thus $ind_x\,(9.7)$, so (9.7) by BI.

9.8 *Thm.* $x \leq y \;\&\; y \leq x \rightarrow x = y$.

Proof. Suppose $x \leq y \;\&\; y \leq x$. Then there exist w and z such that $x + w = y \;\&\; y + z = x$. Hence $x + (w + z) = x + 0$, and $w + z = 0$ by (9.1). Consequently $w \leq 0$, so $w = 0$ and $x = y$. Thus (9.8).

9.9 *Def.* $\mathrm{Max}(x, y) = z \leftrightarrow (x \leq y \;\&\; z = y) \vee (y \leq x \;\&\; z = x)$.

The existence condition holds by (9.7), the uniqueness condition holds by (9.8), and we have $\exists z\, rhs\,(9.9)$: $z \leq x + y$.

By the way, a nonlogical symbol beginning with a capital letter will always be a function symbol (and conversely if the symbol begins with a letter).

9.10 *Def.* $x < y \leftrightarrow x \leq y \;\&\; x \neq y$.

9.11 *Thm.* $y \neq 0 \;\&\; x = y \cdot q_1 + r_1 \;\&\; r_1 < y \;\&\; x = y \cdot q_2 + r_2 \;\&\; r_2 < y \rightarrow q_1 = q_2 \;\&\; r_1 = r_2$.

Proof. Suppose $hyp\,(9.11)$ and suppose $q_1 < q_2$. There exists w such that $q_2 = q_1 + w \;\&\; w \neq 0$. Then $y \cdot q_1 + r_1 = x = y \cdot q_1 + y \cdot w + r_2$, so $r_1 = y \cdot w + r_2$. By (9.5), $y \leq y \cdot w \leq r_1$, which contradicts $r_1 < y$ by (9.8).

Thus $\neg(q_1 < q_2)$, and similarly $\neg(q_2 < q_1)$, so that $q_1 = q_2$ by (9.7). By (9.1), $r_1 = r_2$. Thus (9.11).

9.12 *Thm.* $y \neq 0 \rightarrow \exists q \exists r(x = y \cdot q + r \;\&\; r < y)$.

Proof. We have (9.12)$:q \leq x, r \leq y$. Since $0 = y \cdot 0 + 0 \;\&\; (y \neq 0 \rightarrow 0 < y)$, we have $(9.12)_x[0]$. Suppose $x = y \cdot q + r \;\&\; r < y$. There exists w such that $r + w = y \;\&\; w \neq 0$, so $r + \mathrm{SP}w = y$, $\mathrm{S}r + \mathrm{P}w = y$, and $\mathrm{S}r \leq y$; that is, $\mathrm{S}r < y \vee \mathrm{S}r = y$. Then $\mathrm{S}r < y \rightarrow \mathrm{S}x = y \cdot q + r \;\&\; \mathrm{S}r < y$, and $\mathrm{S}r = y \rightarrow \mathrm{S}x = y \cdot \mathrm{S}q + 0 \;\&\; 0 < y$. Thus ind_x (9.12), so (9.12) by BI.

9.13 *Def.* $\mathrm{Qt}(y, x) = q \leftrightarrow \exists r(x = y \cdot q + r \;\&\; r < y) \vee (y = 0 \;\&\; q = 0)$.

The uniqueness condition holds by (9.11), the existence condition holds by (9.12), and we have $\exists q\, rhs$ (9.13)$:q \leq x, r \leq y$, and similarly for the next defining axiom.

9.14 *Def.* $\mathrm{Rm}(y, x) = r \leftrightarrow \exists q(x = y \cdot q + r \;\&\; r < y) \vee (y = 0 \;\&\; r = 0)$.

9.15 *Def.* $1 = \mathrm{S}0$.

9.16 *Def.* $2 = \mathrm{S}1$.

9.17 *Def.* $3 = \mathrm{S}2$.

9.18 *Def.* $4 = \mathrm{S}3$.

9.19 *Def.* $5 = \mathrm{S}4$.

9.20 *Def.* $6 = \mathrm{S}5$.

9.21 *Def.* $7 = \mathrm{S}6$.

9.22 *Def.* $8 = \mathrm{S}7$.

9.23 *Def.* $9 = \mathrm{S}8$.

9.24 *Def.* $\mathrm{Dec}(x, y) = x \cdot \mathrm{S}9 + y$.

The constants 0, 1, 2, 3, 4, 5, 6, 7, 8, and 9 are called *decimal digits.* If $\mathrm{e}_1, \mathrm{e}_2, \ldots, \mathrm{e}_{\nu-1}, \mathrm{e}_\nu$ are decimal digits, ν is at least 2, and e_1 is not 0, then we write $\mathrm{e}_1\mathrm{e}_2 \ldots \mathrm{e}_{\nu-1}\mathrm{e}_\nu$ for $\mathrm{Dec}(\mathrm{Dec}(\ldots \mathrm{Dec}(\mathrm{e}_1, \mathrm{e}_2) \ldots \mathrm{e}_{\nu-1}), \mathrm{e}_\nu)$. A term that is of this form or is a decimal digit is called a *decimal.* If e_1 and e_2 are decimal digits, one can find decimals a and b such that $\vdash \mathrm{e}_1 + \mathrm{e}_2 = \mathrm{a}$ and $\vdash \mathrm{e}_1 \cdot \mathrm{e}_2 = \mathrm{b}$. The two hundred theorems of this form can conveniently be tabulated as addition and multiplication tables; if decimals $\mathrm{e}_1 \ldots \mathrm{e}_\nu$ and

$\mathrm{e}'_1 \ldots \mathrm{e}'_\mu$ are given, these tables make it easy to find decimals c and d such that $\vdash \mathrm{e}_1 \ldots \mathrm{e}_\nu + \mathrm{e}'_1 \ldots \mathrm{e}'_\mu = \mathrm{c}$ and $\vdash \mathrm{e}_1 \ldots \mathrm{e}_\nu \cdot \mathrm{e}'_1 \ldots \mathrm{e}'_\mu = \mathrm{d}$.

9.25 *Def.* $x \mid y \leftrightarrow \exists z\, x \cdot z = y$.

We have $rhs\,(9.25)\colon z \leq y$.

9.26 *Thm.* $x \mid y \;\&\; y \neq 0 \rightarrow x \leq y$.

Proof. Suppose $hyp\,(9.26)$. There exists z such that $x \cdot z = y$. Then $z \neq 0$, so $x \leq y$ by (9.5). Thus (9.26).

9.27 *Thm.* $a = b + c \;\&\; x \mid a \;\&\; x \mid b \rightarrow x \mid c$.

Proof. Suppose $hyp\,(9.27)$. There exist u and v such that $x \cdot u = a \;\&\; x \cdot v = b$, so $x \cdot u = x \cdot v + c$. Suppose $x = 0$. Then $c = 0 = 0 \cdot 0$ and thus $x = 0 \rightarrow x \mid c$. Suppose $x \neq 0$. Then $v \leq u$, so there exists w such that $v + w = u$. Then $x \cdot v + x \cdot w = x \cdot u$, so $x \cdot w = c$ by (9.1). Thus $x \mid c$ and thus (9.27).

9.28 *Def.* p is a prime $\leftrightarrow p \neq 1 \;\&\; \forall x(x \mid p \rightarrow x = 1 \vee x = p)$.

We have $p \neq 0 \rightarrow rhs\,(9.28)\colon x \leq p$. Also, $2 \mid 0 \;\&\; 2 \neq 1 \;\&\; 2 \neq 0$, so $p = 0 \rightarrow rhs\,(9.28)\colon x \leq 2$. Consequently we have $rhs\,(9.28)\colon x \leq \mathrm{Max}(p, 2)$.

We will frequently introduce predicate symbols containing words, but we use these predicate symbols formally. For example, "$\neg$(0 is a prime)" is a theorem, but "0 is not a prime" is not even a formula. If $***$ is obtained by forming the plural of a noun in —, then we write (omitting the comma if ν is 2) "$\mathrm{a}_1, \ldots,$ and a_ν are $***$" for "a_1 is a — $\;\&\; \cdots \;\&\;$ a_ν is a —".

9.29 *Thm.* p is a prime $\;\&\;$ $p \mid a \cdot b \rightarrow p \mid a \vee p \mid b$.

Proof. Suppose p is a prime $\;\&\;$ $p \mid a \cdot b \;\&\; \neg(p \mid a)$. By (9.12) there exist q and r such that $a = p \cdot q + r \;\&\; r < p$. Also, $0 < r$ since $\neg(p \mid a)$. We have $a \cdot b = p \cdot q \cdot b + r \cdot b$, so $p \mid r \cdot b$ by (9.27). By BLNP there exists r_0 such that $\min_{r_0}(0 < r_0 < p \;\&\; p \mid r_0 \cdot b)$. There exist q_1 and r_1 such that $p = r_0 \cdot q_1 + r_1 \;\&\; r_1 < r$. Since $p \cdot b = r_0 \cdot q_1 \cdot b + r_1 \cdot b$ we have $p \mid r_1 \cdot b$. By the minimality assumption, $r_1 = 0$. Hence $p = r_0 \cdot q_1$ and $r_0 \mid p$. Since $r_0 < p$ and p is a prime, $r_0 = 1$. Therefore $p \mid b$, and thus (9.29).

Chapter 10

Encoding

In the last chapter we copied the usual proofs, observing that only bounded inductions are involved. But now we come to a fork in the road. Arithmetic is too limited unless it can express notions such as finite sums and products, exponentiation, etc. The usual semi-formal treatment of such notions is based on a pun on the word "number", confounding the formal notion of number as a term of a theory with the genetic notion of number used in counting. For example, when one writes

$$\sum_{i=1}^{n} \mathrm{f}(i) = \mathrm{f}(1) + \cdots + \mathrm{f}(n),$$

one is trying simultaneously to use n as a term of a theory and to speak of n terms of that theory. Gödel overcame this difficulty by finding a way to represent recursive functions within Peano Arithmetic. The first requirement is for a means of encoding a finite set of numbers by a single number. Gödel does this by means of his beta function (see [Gö] or [Sh,§6.4]). This method is impredicative; in fact, it relies precisely on the induction giving $\pi(n)$ discussed in the second example of Chapter 1. Mostowski, Robinson, and Tarski showed (see [MRT,pp.56–59]) that recursive functions are representable in Q, but this does not meet our needs for two reasons: first, the methods are impredicative (theirs is a result about Q proved in a stronger theory) and second, recursive functions are represented only extensionally in Q, whereas we desire, where possible, an intensional representation (see [Fe]) allowing us to prove properties of the functions within the theory. Therefore we will seek a different route.

The encoding method that we will use is this: take a number a, write it in base four notation, and consider the set of *all* numbers whose base two representation occurs in the base four representation of a immediately preceded and followed by the base four digit 2; then we regard a as encoding this finite set of numbers. For example, the number a whose base four representation is 202112 (i.e., 2198 in decimal notation) encodes the pair of numbers whose base two representations are 0 and 11 (i.e., 0 and 3 in decimal notation). We need to show that this encoding can be expressed within our theory. This is not immediately clear; for example, one can speak of the first, second, or third binary digit of a number (counting from the right), but how can one speak of its kth binary digit? We will get around this problem by speaking instead of the binary digit in the q's place, where q is a power of two.

10.1 *Def.* q is a power of two $\leftrightarrow \forall p(p$ is a prime $\& \ p \mid q \rightarrow p = 2)$.

We have $q \neq 0 \rightarrow rhs\,(10.1){:}\,p \leq q$. Also, 3 is a prime $\ \& \ 3 \mid 0 \ \& \ 3 \neq 2$, so that $rhs\,(10.1){:}\,p \leq \text{Max}(q, 3)$.

10.2 *Def.* q is a power of four $\leftrightarrow q$ is a power of two $\& \ \exists r \ r \cdot r = q$.

We have $rhs\,(10.2){:}\,r \leq q$. Let us agree to stop giving bounds on formulas when they are obvious.

10.3 *Def.* $\text{power}(b, q) \leftrightarrow (b = 2 \ \& \ q$ is a power of two$) \vee (b = 4 \ \& \ q$ is a power of four$)$.

10.4 *Thm.* $\text{power}(b, q_1) \ \& \ \text{power}(b, q_2) \rightarrow \text{power}(b, q_1 \cdot q_2)$.

Proof. Suppose $hyp\,(10.4)$, so that $b = 2 \vee b = 4$, and suppose p is a prime $\& \ p \mid q_1 \cdot q_2$. By (9.29), $p \mid q_1 \vee p \mid q_2$. But q_1 and q_2 are powers of two, so $p = 2$. Thus $q_1 \cdot q_2$ is a power of two, and therefore $b = 2 \rightarrow con\,(10.2)$. Suppose $b = 4$. Then there exist r_1 and r_2 such that $r_1 \cdot r_1 = q_1 \ \& \ r_2 \cdot r_2 = q_2$, so $(r_1 \cdot r_2) \cdot (r_1 \cdot r_2) = q_1 \cdot q_2$ and thus $b = 4 \rightarrow con\,(10.4)$. Thus (10.4).

10.5 *Thm.* $\text{power}(b, q) \ \& \ q \neq 1 \rightarrow \text{power}(b, q/b)$.

Proof. Suppose $hyp\,(10.5)$ and observe that $\neg(0$ is a power of two$)$, so that $1 < q \ \& \ q \mid q$, so by BLNP there exists p such that $\min_p(1 < p \ \& \ p \mid q)$. Suppose $x \mid p$. Then $x \mid q$, so by the minimality of p we have $x = 1 \vee x = p$.

Thus p is a prime, and $p \mid q$, so $p = 2$. That is, $2 \mid q$, so that $q = (q/2) \cdot 2$. Suppose y is a prime & $y \mid (q/2)$. Then $y \mid q$, so $y = 2$. Thus $q/2$ is a power of two, and consequently $b = 2 \to con\,(10.5)$. Suppose $b = 4$. Then there exists r such that $r \cdot r = q$. Since 2 is a prime and $2 \mid q$, we have $2 \mid r$ by (9.29), so $4 \cdot (r/2) \cdot (r/2) = q$. Hence $4 \mid q$, and $(r/2) \cdot (r/2) = q/4$. Suppose z is a prime & $z \mid (q/4)$. Then $z \mid q$, so $z = 2$. Thus $q/4$ is a power of four, and thus $b = 4 \to con\,(10.5)$. Thus (10.5).

10.6 *Thm.* $\mathrm{power}(b, q_1)$ & $\mathrm{power}(b, q_2)$ & $q_1 \leq q_2 \to \mathrm{power}(b, q_2/q_1)$.

Proof. Suppose $\exists q_1 \exists q_2 \neg(10.6)$. By BLNP there exist minimal such q_1 and q_2. Then $q_1 \neq 1$. By (10.5), $\mathrm{power}(b, q_1/b)$, so by the minimality assumption we have $\mathrm{power}(b, q_2/(q_1/b))$. That is, $\mathrm{power}(b, (q_2 \cdot b)/q_1)$. Since $q_1 \leq q_2$, we have $(q_2 \cdot b)/q_1 \neq 1$. By (10.5) again, $\mathrm{power}(b, ((q_2 \cdot b)/q_1)/b)$. That is, $\mathrm{power}(b, q_2/q_1)$, which is a contradiction. Thus (10.6).

10.7 *Thm.* $\mathrm{power}(b, q_1)$ & $q_1 \leq x < q_1 \cdot b$ & $\mathrm{power}(b, q_2)$ & $q_2 \leq x < q_2 \cdot b \to q_1 = q_2$.

Proof. Suppose $hyp\,(10.7)$ & $q_1 \leq q_2$. We have $q_2 < q_1 \cdot b$, so $q_2/q_1 < b$ and $\neg(b \mid (q_2/q_1))$. By (10.5), $q_2/q_1 = 1$, so $q_1 = q_2$. Thus $q_1 \leq q_2 \to (10.7)$, and therefore $q_2 \leq q_1 \to (10.7)$, so that (10.7).

10.8 *Def.* $|x|_b = q \leftrightarrow \mathrm{power}(b, q)$ & $x \neq 0$ & $q \leq x < q \cdot b$, otherwise $q = 1$.

The formulas in what follows will be easier to read if we bear in mind that, for a base b and q a power of b, the base b representation of x is that of $\mathrm{Qt}(q, x)$ down to and including the q's place, followed by that of $\mathrm{Rm}(q, x)$. Using these scissors we can snip out any desired portion of the representation of x. For example, the base b digit in the q's place of x is $\mathrm{Rm}(b, \mathrm{Qt}(q, x))$. It is intuitive that repeated snipping does not change the value of any digit (though perhaps its location may change), and the next theorem is a result of this sort.

10.9 *Thm.* $\mathrm{power}(b, q_1)$ & $\mathrm{power}(b, q_2)$ & $q_1 < q_2 \to \mathrm{Rm}(b, \mathrm{Qt}(q_1, x)) = \mathrm{Rm}(b, \mathrm{Qt}(q_1, \mathrm{Rm}(q_2, x)))$.

Proof. Suppose $hyp\,(10.9)$. We claim that

1. $x = \mathrm{Qt}(q_2, x) \cdot q_2 + \mathrm{Rm}(q_2, x)$,

2. $\mathrm{Rm}(q_2, x) = \mathrm{Qt}(q_1, \mathrm{Rm}(q_2, x)) \cdot q_1 + \mathrm{Rm}(q_1, \mathrm{Rm}(q_2, x))$,

3. $x = \mathrm{Qt}(q_1, x) \cdot q_1 + \mathrm{Rm}(q_1, x)$,

4. $x = (\mathrm{Qt}(q_2, x) \cdot (q_2/q_1) + \mathrm{Qt}(q_1, \mathrm{Rm}(q_2, x))) \cdot q_1 + \mathrm{Rm}(q_1, \mathrm{Rm}(q_2, x))$,

5. $\mathrm{Qt}(q_1, x) = \mathrm{Qt}(q_2, x) \cdot (q_2/q_1) + \mathrm{Qt}(q_1, \mathrm{Rm}(q_2, x))$.

We have (1), (2), and (3) directly from the defining axioms for Qt and Rm, (4) is a consequence of (1) and (2), and (5) is a consequence of (3) and (4). But $q_1 < q_2$, so $b \mid (q_2/q_1)$ and (5) gives *con* (10.9). Thus (10.9). □

The next two theorems show that the first significant digit of x occurs in the $|x|_b$'s place.

10.10 *Thm.* $\mathrm{power}(b, q)\ \&\ |x|_b < q \rightarrow \mathrm{Rm}(b, \mathrm{Qt}(q, x)) = 0$.

Proof. Suppose *hyp* (10.10). Then $x < q$, so $\mathrm{Qt}(q, x) = 0$ and therefore $\mathrm{Rm}(b, \mathrm{Qt}(q, x)) = 0$. Thus (10.10).

10.11 *Thm.* $x \neq 0\ \&\ (b = 2 \vee b = 4) \rightarrow 0 < \mathrm{Rm}(b, \mathrm{Qt}(|x|_b, x)) = \mathrm{Qt}(|x|_b, x) < b$.

Proof. Suppose *hyp* (10.11). Then $|x|_b \leq x < |x|_b \cdot b$, so that we have $0 < \mathrm{Qt}(|x|_b, x) < b$, and hence *con* (10.11). Thus (10.11). □

Next we show that two numbers with the same base b representation are equal. We do this by the method of infinite descent (predicatively), snipping off the first significant digit to get a pair of smaller numbers with the same base b representation.

10.12 *Thm.* $(b = 2 \vee b = 4)\ \&\ \forall q(\mathrm{power}(b, q) \rightarrow \mathrm{Rm}(b, \mathrm{Qt}(q, x_1)) = \mathrm{Rm}(b, \mathrm{Qt}(q_1, x_2))) \rightarrow x_1 = x_2$.

Proof. We have (10.12):$q \leq \mathrm{Max}(|x_1|_b, |x_2|_b)$ by (10.10). Suppose $\exists x_1 \exists x_2 \neg$(10.12). By BLNP there exist minimal such x_1 and x_2. Suppose $x_1 = 0$. Then $\mathrm{Rm}(b, \mathrm{Qt}(|x_2|_b, x_2)) = 0$, so $x_2 = 0$ by (10.11). Hence $x_1 = x_2$, a contradiction, and thus $x_1 \neq 0$. Similarly, $x_2 \neq 0$. By (10.10) and (10.11), $|x_1|_b = |x_2|_b$. Let $q_0 = |x_1|_b$, let $y_1 = \mathrm{Rm}(q_0, x_1)$, and let $y_2 = \mathrm{Rm}(q_0, x_2)$. Then $y_1 < x_1$ and $y_2 < x_2$, and

$$x_1 = \mathrm{Qt}(q_0, x_1) \cdot q_0 + y_1,$$

$$x_2 = \mathrm{Qt}(q_0, x_2) \cdot q_0 + y_2.$$

We have $\mathrm{Rm}(b, \mathrm{Qt}(q_0, x_1)) = \mathrm{Rm}(b, \mathrm{Qt}(q_0, x_2))$, so by (10.11) we obtain $\mathrm{Qt}(q_0, x_1) = \mathrm{Qt}(q_0, x_2)$. Since $x_1 \neq x_2$ we have $y_1 \neq y_2$. We use the abbreviation

α for $scope_{\forall q}(10.12)_{x_1 x_2}[y_1 y_2]$.

By (10.10), $q_0 \le q \to \alpha$ and by (10.9), $q < q_0 \to \alpha$. Therefore $\forall q\, \alpha$, which contradicts the minimality assumption. Thus (10.12). □

The $q \cdot q$'s place in base four corresponds to the q's place in base two. Next we show that for any number x there is a number z whose base four representation is the same as the base two representation of x.

10.13 *Thm.* $\exists z \forall q(q$ is a power of two $\to$ $\mathrm{Rm}(2, \mathrm{Qt}(q, x)) = \mathrm{Rm}(4, \mathrm{Qt}(q \cdot q, z)))$.

Proof. We have (10.13): $z \le |x|_2 \cdot |x|_2 \cdot 4, q \le |x|_2$ by (10.10). Suppose $\exists x \neg(10.13)$. By BLNP there exists a minimal such x. Suppose $x = 0$ and let $z = 0$. Then (10.13), a contradiction, and thus $x \ne 0$. Let $y = \mathrm{Rm}(|x|_2, x)$, so that $y < x$. By the minimality of x, there exists w such that

$$\forall q(q \text{ is a power of two } \to \mathrm{Rm}(2, \mathrm{Qt}(q, y)) = \mathrm{Rm}(4, \mathrm{Qt}(q \cdot q, w))).$$

Let $z = |x|_2 \cdot |x|_2 + w$, and write α for $scope_{\forall q}(10.13)$. By (10.9), $q < |x|_2 \to \alpha$ and by (10.10), $|x|_2 < q \to \alpha$. By (10.11) we have $0 < \mathrm{Qt}(|x|_2, x) < 2$, so $\mathrm{Rm}(2, \mathrm{Qt}(|x|_2, x)) = 1$. But $\mathrm{Qt}(|x|_2, y) = 0$, so

$$\mathrm{Rm}(4, \mathrm{Qt}(|x|_2 \cdot |x|_2, w)) = \mathrm{Rm}(2, \mathrm{Qt}(|x|_2, y)) = 0.$$

Hence $\mathrm{Rm}(4, \mathrm{Qt}(|x|_2 \cdot |x|_2, z)) = 1$, so that $q = |x|_2 \to \alpha$. Consequently $\forall q\, \alpha$, and so (10.13), a contradiction. Thus (10.13).

10.14 *Def.* $\mathrm{Enc}\, x = z \leftrightarrow \forall q(q$ is a power of two $\to \mathrm{Rm}(2, \mathrm{Qt}(q, x)) = \mathrm{Rm}(4, \mathrm{Qt}(q \cdot q, z)))$.

The uniqueness condition follows from (10.12), the existence condition is (10.13), and we have $\exists z\, rhs\,(10.14)$: $z \le |x|_2 \cdot |x|_2 \cdot 4, q \le |x|_2$. This is the function symbol that we need for encoding.

10.15 *Thm.* $\mathrm{Enc}\, x_1 = \mathrm{Enc}\, x_2 \to x_1 = x_2$.

Proof. By (10.12).

10.16 *Def.* $\mathrm{enc}(q_1, q_2, x, a) \leftrightarrow$ q_1 and q_2 are powers of four & $\mathrm{Rm}(4, \mathrm{Qt}(q_1, a)) = \mathrm{Rm}(4, \mathrm{Qt}(q_2, a)) = 2$ & $q_2 = 4 \cdot |\mathrm{Enc}\, x|_4 \cdot 4 \cdot q_1$ & $\mathrm{Rm}(q_2/(q_1 \cdot 4), \mathrm{Qt}(q_1 \cdot 4, a)) = \mathrm{Enc}\, x$.

10.17 *Def.* $x \in a \leftrightarrow \exists q_1 \exists q_2\, \mathrm{enc}(q_1, q_2, x, a)$.

We have *rhs* (10.17): $q_1 \le a, q_2 \le a$.

We need another scissors lemma, to verify that if two different numbers x and y are encoded in a, then they are encoded in a non-overlapping way. This is used to prove that if x is encoded in a, then there is a b, smaller by at least a factor of 2, such that every other number encoded in a is encoded in b.

10.18 *Thm.* $\mathrm{power}(b, q_1)$ & $\mathrm{power}(b, q)$ & $\mathrm{power}(b, q_2)$ & $q_1 \le q < q_2 \to \mathrm{Rm}(b, \mathrm{Qt}(q/q_1, \mathrm{Rm}(q_2, \mathrm{Qt}(q_1, a)))) = \mathrm{Rm}(b, \mathrm{Qt}(q, a))$.

Proof. Suppose *hyp* (10.18). We claim that

1. $a = \mathrm{Qt}(q_1, a) \cdot q_1 + \mathrm{Rm}(q_1, a)$,

2. $\mathrm{Qt}(q_1, a) = \mathrm{Qt}(q_2, \mathrm{Qt}(q_1, a)) \cdot q_2 + \mathrm{Rm}(q_2, \mathrm{Qt}(q_1, a))$,

3. $\mathrm{Qt}(q/q_1, \mathrm{Qt}(q_1, a)) = \mathrm{Qt}(q_2, \mathrm{Qt}(q_1, a)) \cdot ((q_2 \cdot q_1)/q) + \mathrm{Qt}(q/q_1, \mathrm{Rm}(q_2, \mathrm{Qt}(q_1, a)))$,

4. $\mathrm{Qt}(q/q_1, \mathrm{Qt}(q_1, a)) = \mathrm{Qt}(q, \mathrm{Qt}(q_1, a) \cdot q_1)$,

5. $\mathrm{Qt}(q, \mathrm{Qt}(q_1, a) \cdot q_1) = \mathrm{Qt}(q, a)$,

6. $\mathrm{Qt}(q_2, \mathrm{Qt}(q_1, a)) \cdot ((q_2 \cdot q_1)/q) + \mathrm{Qt}(q/q_1, \mathrm{Rm}(q_2, \mathrm{Qt}(q_1, a))) = \mathrm{Qt}(q_1, a)$.

We have (1) and (2) directly. Since $(q/q_1) \mid q_2$, (3) follows from (2), and (4) is obvious. Since $q_1 \le q$ and $\mathrm{Rm}(q_1, a) < q_1$, (5) follows from (1). By (3), (4), and (5) we have (6). But $b \mid ((q_2 \cdot q_1)/q)$ since $q < q_2$, so *con* (10.18). Thus (10.18).

10.19 *Thm.* $\mathrm{enc}(q_1, q_2, x, a,)$ & $\mathrm{enc}(q_3, q_4, y, a)$ & $x \ne y \to$
$q_1 < q_2 \le q_3 < q_4 \;\vee\; q_3 < q_4 \le q_1 < q_2$.

Proof. Suppose *hyp* (10.19), and suppose $(q = q_3 \vee q = q_4)$ & $q_1 < q < q_2$. We have $(\mathrm{q_2}/(\mathrm{q_1} \cdot 4), \mathrm{Qt}(\mathrm{q_1} \cdot 4, \mathrm{a})) = \mathrm{Enc}\,\mathrm{x}$, so by (10.18) we obtain

$$\mathrm{Rm}(4, \mathrm{Qt}(q/(q_1 \cdot 4), \mathrm{Enc}\, x)) = \mathrm{Rm}(4, \mathrm{Qt}(q, a)) = 2.$$

But since $q/(q_1 \cdot 4)$ is a power of four, there exists r such that r is a power of two and $r \cdot r = q/(q_1 \cdot 4)$, so

$$\mathrm{Rm}(4, \mathrm{Qt}(q/(q_1 \cdot 4), \mathrm{Enc}\, x)) = \mathrm{Rm}(2, q(r, x)) < 2,$$

a contradiction. Thus

1. $\neg(q_1 < q_3 < q_2)$ & $\neg(q_1 < q_4 < q_2)$,

and similarly

2. $\neg(q_3 < q_1 < q_4)$ & $\neg(q_3 < q_2 < q_4)$.

Suppose $q_1 = q_3$, so that $q_3 < q_2$ and $q_1 < q_4$. By (1), $q_2 \leq q_4$ and by (2), $q_2 = q_4$. Consequently $\operatorname{Enc} x = \operatorname{Enc} y$ and so $x = y$, a contradiction. Thus $q_1 \neq q_3$. Suppose $q_1 < q_3$, so that $q_1 < q_3 < q_4$. By (1) we have $q_2 \leq q_3$, and thus $q_1 < q_3 \rightarrow$ *con* (10.19). Similarly, $q_3 < q_1 \rightarrow$ *con* (10.19), and thus (10.19).

10.20 *Thm.* $x \in a \rightarrow \exists b(2 \cdot b < a \;\&\; \forall y(y \neq x \;\&\; y \in a \rightarrow y \in b))$.

Proof. Suppose $x \in a$, so there exist q_1 and q_2 such that $\operatorname{enc}(q_1, q_2, x, a)$. Let $b = \operatorname{Qt}(q_2, a) \cdot q_1 \cdot 4 + \operatorname{Rm}(q_1 \cdot 4, a)$. (That is, b is obtained from a by deleting $\operatorname{Enc} x$.) Since $q_1 \cdot 4 \cdot 4 \leq q_2$, $\operatorname{Rm}(q_1 \cdot 4, a) < q_1 \cdot 4$, and $\operatorname{Qt}(q_2, a) \cdot q_2 \leq a$, we have $4 \cdot b < 2 \cdot a$, so that $2 \cdot b < a$. Suppose 1: $y \neq x \;\&\; y \in a$, so there exist q_3 and q_4 such that $\operatorname{enc}(q_3, q_4, y, a)$. By (10.19), $q_1 < q_2 \leq q_3 < q_4 \vee q_3 < q_4 \leq q_1 < q_2$. Suppose 2: $q_1 < q_2 \leq q_3 < q_4$. Then $\operatorname{enc}(q_3/|\operatorname{Enc} x|_4, q_4/|\operatorname{Enc} x|_4, y, b)$, and thus (2) $\rightarrow y \in b$. Suppose 3: $q_3 < q_4 \leq q_1 < q_2$. Then $\operatorname{enc}(q_3, q_4, y, b)$, and thus (3) $\rightarrow y \in b$. Therefore $y \in b$, and thus (1) $\rightarrow y \in b$. Thus (10.20).

10.21 *Thm.* $x \in a \rightarrow x < a$.

Proof. Suppose $x \in a$. Then $\operatorname{Enc} x < a$. But $0 = \operatorname{Enc} 0$, $1 = \operatorname{Enc} 1$, and $2 \leq x \rightarrow x < |x|_2 \cdot 2 \leq |x|_2 \cdot |x|_2 = |\operatorname{Enc} x|_4 \leq \operatorname{Enc} x$, so $x \leq \operatorname{Enc} x < a$. Thus (10.21).

10.22 *Def.* a is a set $\leftrightarrow \neg\exists b(b < a \;\&\; \forall x(x \in a \rightarrow x \in b))$.

Chapter 11

Bounded separation and minimum

Let Q_2' be the current theory; that is, the extension of Q_2 obtained by adjoining the defining axioms up to the present.

Metatheorem 11.1 Let T *be an extension of* Q_2', *let* A *be a bounded formula of* T, *and let* x, y, *and* z *be distinct variables such that* z *does not occur in* A. *Then*

$$\text{BSD. } \{\mathrm{x} \in \mathrm{y} : \mathrm{A}\} = \mathrm{z} \leftrightarrow \min_{\mathrm{z}} \forall \mathrm{x}(\mathrm{x} \in \mathrm{y} \ \&\ \mathrm{A} \to \mathrm{x} \in \mathrm{z})$$

is the defining axiom of a bounded function symbol. (The variables in the term $\{\mathrm{x} \in \mathrm{y} : \mathrm{A}\}$ *are* y *and the variables distinct from* x *that occur free in* A.*) The following is a theorem of* T[(BSD)]:

$$\begin{aligned}\text{BS. } & \{\mathrm{x} \in \mathrm{y} : \mathrm{A}\} \text{ is a set } \&\ \{\mathrm{x} \in \mathrm{y} : \mathrm{A}\} \leq \mathrm{y}\ \& \\ & (\mathrm{x} \in \{\mathrm{x} \in \mathrm{y} : \mathrm{A}\} \leftrightarrow \mathrm{x} \in \mathrm{y}\ \&\ \mathrm{A}).\end{aligned}$$

Demonstration. The uniqueness condition for (BSD) holds by (9.8) and the definition of min in Chapter 8. By (10.21),

$$\vdash_{\mathrm{T}} \forall \mathrm{x}(\mathrm{x} \in \mathrm{y} \ \&\ \mathrm{A} \to \mathrm{x} \in \mathrm{z}) \mathpunct{:} \mathrm{x} \leq \mathrm{y},$$

and of course

$$\vdash_T \exists z \forall x(x \in y \ \& \ A \to x \in z).$$

Therefore, by BLNP we have the existence condition for (BSD), and

$$\vdash_T \exists z \, rhs \, (\text{BSD}) : z \le y, x \le y.$$

Hence (BSD) is the defining axiom of a bounded function symbol. To prove (BS) in T[(BSD)], we argue as follows.

We have $\{x \in y : A\}$ is a set & $\{x \in y : A\} \le y$ from the defining axioms (10.22) and (BSD). Let $z = \{x \in y : A\}$. Clearly

1. $x \in y \ \& \ A \to x \in z.$

Suppose $x \in z \ \& \ \neg(x \in y \ \& \ A)$. By (10.20) there exists z_1 such that

$$z_1 < z \ \& \ \forall x_1(x_1 \ne x \ \& \ x_1 \in z \to x_1 \in z_1).$$

(Here z_1 and x_1 are distinct, and are distinct from x, y, z, and all variables occurring in A). Suppose $x_1 \in y \ \& \ A_x[x_1]$. Then $x_1 \ne x$, and $x_1 \in z$ by (1), so $x_1 \in z_1$. Thus

$$\forall x_1(x_1 \in y \ \& \ A_x[x_1] \to x_1 \in z_1),$$

which contradicts the minimality of z. Thus (BS). □

Given a formula A and a variable x, let y be the first variable in alphabetical order distinct from x and all variables in A. Then we write

$$\max\nolimits_x A \quad \text{for} \quad A \ \& \ \neg\exists y(x < y \ \& \ A_x[y]).$$

(We could also define $\max_{x_1 \dots x_\nu} A$ in the obvious way, but it does not seem likely that we would ever have occasion to use it.)

Metatheorem 11.2 Let T *be an extension of* Q_2', *let* A *be a formula of* T, *and let* x, y, *and* z *be distinct variables such that* z *does not occur in* A. *Then*

MIND. $\text{Min}\, x(x \le y \ \& \ A) = z \leftrightarrow \min_z(z \le y \ \& \ A_x[z])$,
otherwise $z = 0$,

MAXD. $\text{Max}\, x(x \le y \ \& \ A) = z \leftrightarrow \max_z(z \le y \ \& \ A_x[z])$,
otherwise $z = 0$

are defining axioms of bounded function symbols. (The variables in these terms are y *and all variables distinct from* x *that occur free in* A.*) The following are theorems of* T[(MIND)] *and* T[(MAXD)] *respectively*:

$$\text{MIN. } x \leq y \ \& \ A \to \text{Min}\, x(x \leq y \ \& \ A) \leq x \ \& \ A_x[\text{Min}\, x(x \leq y \ \& \ A)],$$

$$\text{MAX. } x \leq y \ \& \ A \to x \leq \text{Max}\, x(x \leq y \ \& \ A) \ \& \ A_x[\text{Max}\, x(x \leq y \ \& \ A)].$$

Demonstration. The uniqueness conditions hold by (9.7) and the definitions of min and max. We have

$$\vdash_T \exists z \ \mathit{rhs}\,(\text{MIND}){:}\, z \leq y,$$

and similarly for (MAXD), so these are defining axioms of bounded function symbols.

To prove (MIN) in T[(MIND)] we argue as follows. Suppose $x \leq y \ \& \ A$. By BLNP there exists z such that

$$\min_z(z \leq y \ \& \ A_x[z]),$$

so that $\text{Min}\, x(x \leq y \ \& \ A) = z$. Then $z \leq x \ \& \ A_x[z]$. Thus (MIN).

To prove (MAX) in T[(MAXD)] we argue as follows. Suppose $x < y \ \& \ A$. Then $y - x \leq y \ \& \ A_x[y - (y - x)]$. By BLNP there exists z such that

$$\min_z(z \leq y \ \& \ A_x[y - z]),$$

so that $y - z = \text{Max}\, x(x \leq y \ \& \ A)$. By MIN, $z \leq y - x \ \& \ A_x[y - z]$. Thus (MAX). □

Whenever a term of the form $\text{Min}\, x(x \leq a \ \& \ A)$, $\text{Max}\, x(x \leq a \ \& \ A)$, or $\{x \in a : A\}$ occurs, it is understood that the corresponding defining axiom has been adjoined to the theory.

If all occurrences of Min x in B are in the term $\text{Min}\, x(x \leq a \ \& \ A)$, we write

$$\mathit{scope}_{\text{Min}\, x}B \ \text{ for } \ x \leq a \ \& \ A,$$

and similarly with Min x replaced by Max x. We sometimes omit x and write $\mathit{scope}_{\text{Min}}B$.

Chapter 12

Sets and functions

Although conceptually simple, the encoding procedure is laborious. In this chapter we will develop an elementary theory (in the mathematical sense of a collection of theorems) of certain finite sets of numbers, and afterwards we can hopefully forget the details of the encoding procedure.

12.1 *Thm.* 0 is a set $\;\&\; \neg(x \in 0)$.

Proof. Clearly 0 is a set. By(10.21), $\neg(x \in 0)$.

12.2 *Thm.* a and b are sets $\;\&\; \forall x(x \in a \leftrightarrow x \in b) \rightarrow a = b$.

Proof. Suppose hyp(12.2). Then $a \leq b \leq a$. Thus (12.2).

12.3 *Def.* $\{x\} = \{y \in 2 \cdot |\mathrm{Enc}\, x|_4 \cdot 4 \cdot 4 + \mathrm{Enc}\, x \cdot 4 + 2 : y = x\}$.

12.4 *Thm.* $\{x\}$ is a set $\;\&\; (y \in \{x\} \leftrightarrow y = x)$.

Proof. By BS, $\{x\}$ is a set $\;\&\; (y \in \{x\} \rightarrow y = x)$. We have

$$\mathrm{enc}(1, |\mathrm{Enc}\, x|_4 \cdot 4 \cdot 4, x, 2 \cdot |\mathrm{Enc}\, x|_4 \cdot 4 \cdot 4 + \mathrm{Enc}\, x \cdot 4 + 2),$$

so $x \in \{x\}$.

12.5 *Def.* $a \cup b = \{x \in a \cdot |b|_4 \cdot 4 + b : x \in a \vee x \in b\}$.

12.6 *Thm.* $a \cup b$ is a set $\;\&\; (x \in a \cup b \leftrightarrow x \in a \vee x \in b)$.

Proof. By BS, $a \cup b$ is a set $\;\&\; (x \in a \cup b \rightarrow x \in a \vee x \in b)$. Let $c = a \cdot |b|_4 \cdot 4 + b$. (We do not claim that c is a set.) Suppose $x \in a$.

Then there exist q_1 and q_2 such that $\mathrm{enc}(q_1, q_2, x, a)$. Let $q_3 = q_1 \cdot |b|_4 \cdot 4$ and let $q_4 = q_2 \cdot |b|_4 \cdot 4$. Then $\mathrm{enc}(q_3, q_4, x, c)$, so that $x \in c$ and $x \in a \cup b$. Thus $x \in a \rightarrow x \in a \cup b$. Suppose $x \in b$. There exist q_1 and q_2 such that $\mathrm{enc}(q_1, q_2, x, b)$, and we also have $\mathrm{enc}(q_1, q_2, x, c)$. Thus $x \in b \rightarrow x \in a \cup b$, and so (12.6).

12.7 *Def.* $a \cap b = \{x \in a : x \in b\}$.

12.8 *Def.* $a \subseteq b \leftrightarrow \forall x(x \in a \rightarrow x \in b)$.

12.9 *Thm.* a is a set $\ \&\ a \subseteq b \rightarrow a \leq b$.

Proof. Suppose $hyp(12.9)$. Then it follows from BS that we have $a = \{x \in a : x \in a\} = \{x \in b : x \in a\} \leq b$. Thus (12.9).

12.10 *Def.* $\langle x, y \rangle = (x + y) \cdot (x + y) + y$.

Notice that the ordered pair $\langle x, y \rangle$ need not be a set, but it has the following crucial property:

12.11 *Thm.* $\langle x_1, y_1 \rangle = \langle x_2, y_2 \rangle \rightarrow x_1 = x_2 \ \&\ y_1 = y_2$.

Proof. Suppose $hyp(12.11)$ and suppose $x_1 + y_1 < x_2 + y_2$. There exists z such that $x_1 + y_1 + z = x_2 + y_2 \ \&\ z \neq 0$. Then

$$\langle x_2, y_2 \rangle = (x_1 + y_1 + z) \cdot (x_1 + y_1 + z) + y_2 =$$
$$(x_1 + y_1) \cdot (x_1 + y_1) + 2 \cdot z \cdot y_1 + 2 \cdot z \cdot x_1 + z \cdot z + y_2,$$

so $\langle x_1, y_1 \rangle < \langle x_2, y_2 \rangle$, a contradiction. Thus $x_2 + y_2 \leq x_1 + y_1$, and similarly $x_1 + y_1 \leq x_2 + y_2$, so that $x_1 + y_1 = x_2 + y_2$. By $hyp(12.11)$ and (9.1) we have $y_1 = y_2$, and so $x_1 = x_2$. Thus (12.11).

12.12 *Thm.* $x \leq \langle x, y \rangle \ \&\ y \leq \langle x, y \rangle$.

Proof. Clearly $y \leq \langle x, y \rangle$. Also, $x \leq x \cdot x \leq \langle x, y \rangle$.

12.13 *Def.* f is a function $\leftrightarrow f$ is a set $\ \&\ \forall w(w \in f \rightarrow \exists x \exists y\, \langle x, y \rangle = w)$ $\&\ \forall x \forall y_1 \forall y_2(\langle x, y_1 \rangle \in f \ \&\ \langle x, y_2 \rangle \in f \rightarrow y_1 = y_2)$.

12.14 *Def.* $f(x) = y \leftrightarrow f$ is a function $\ \&\ \langle x, y \rangle \in f$, otherwise $y = 0$.

The uniqueness condition follows from (12.13). In addition, we have $\exists y\, rhs(12.14)\colon y \leq f$, so (12.14) is the defining axiom of a bounded function symbol. Notice that it is a binary function symbol; $f(x)$ is a term with two variables, f and x.

12.15 *Def.* x is in the domain of $f \leftrightarrow f$ is a function & $\exists y\, \langle x, y\rangle \in f$.

12.16 *Def.* y is in the range of $f \leftrightarrow f$ is a function & $\exists x\, \langle x, y\rangle \in f$.

12.17 *Thm.* f and g are functions & $f(j) = g(j)$ &
$\forall i(i$ is in the domain of $f \leftrightarrow i$ is in the domain of $g \leftrightarrow j \le i \le k)$ &
$\forall i(j \le i < k\ \&\ f(i) = g(i) \rightarrow f(i+1) = g(i+1)) \rightarrow f = g$.

Proof. Suppose *hyp*(12.17). By (12.2) we need only establish that $\forall w(w \in f \leftrightarrow w \in g)$, so we need only show 1: $\forall i\, f(i) = g(i)$. Suppose $\neg$(1). By BLNP there exists i such that $\min_i f(i) \neq g(i)$. Clearly $j \le i \le k$. Since $f(j) = g(j)$, we have $i \neq j$. Therefore $j \le i - 1 < k$, and by the minimality of i we have $f(i-1) = g(i-1)$. Hence $f(i) = g(i)$, a contradiction, and thus (1). Thus (12.17). □

Let us try to construct bounded function symbols Dom and Ran to express the domain and range of a function. It is intuitive from the encoding procedure that the domain or range of a function is smaller than the function, but to establish this we need to refer to the details of the encoding procedure—let us hope for the last time.

12.18 *Thm.* $x \le y \rightarrow \mathrm{Enc}\, x \le \mathrm{Enc}\, y$.

Proof. Suppose $\exists x \exists y \neg$(12.18). By BLNP there exist minimal such x and y. Clearly $x \neq 0$ and $y \neq 0$. Let $u = \mathrm{Rm}(|x|_2, x)$ and let $v = \mathrm{Rm}(|y|_2, y)$. By (10.11), $\mathrm{Qt}(|x|_2, x) = 1 = \mathrm{Qt}(|y|_2, y)$, so $x = |x|_2 + u$ and $y = |y|_2 + v$, where $u < |x|_2$ and $v < |y|_2$. We have $\mathrm{Enc}\, x = |x|_2 \cdot |x|_2 - \mathrm{Enc}\, u$ and $\mathrm{Enc}\, y = |y|_2 \cdot |y|_2 + \mathrm{Enc}\, v$. Also, $|x|_2 \le |y|_2$. Suppose $|x|_2 < |y|_2$. Then $\mathrm{Enc}\, x < \mathrm{Enc}\, y$, a contradiction, and thus $|x|_2 = |y|_2$. Consequently $u \le v$. By the minimality assumption, $\mathrm{Enc}\, u \le \mathrm{Enc}\, v$, so that $\mathrm{Enc}\, x \le \mathrm{Enc}\, y$, a contradiction. Thus (12.18).

$\mathrm{Rm}(q_1 \cdot 4, b)$. Then $b_0 < b$. Suppose $x_0 \in a_0$. Then $x_0 \neq x$, so $f(x_0) \neq f(x)$. Since $f(x_0) \in b$, there exist q_3 and q_4 such that $\mathrm{enc}(q_3, q_4, f(x_0), b)$. By (10.19) and the definition of q_1 we have $q_3 < q_4 \leq q_1 < q_2$. Consequently $f(x_0) \in b_0$. Thus, by the minimality assumption, $a_0 \leq b_0$. Observe that

$$2 \cdot |\mathrm{Enc}\, f(x)|_4 \cdot q_1 \cdot 4 \cdot 4 + \mathrm{Enc}\, f(x) \cdot q_1 \cdot 4 \cdot b_0 \leq b,$$

and let

$$a_1 = 2 \cdot |\mathrm{Enc}\, x|_4 \cdot q_1 \cdot 4 \cdot 4 + \mathrm{Enc}\, x \cdot q_1 \cdot 4 + a_0.$$

Since $x \leq f(x)$ we have $a_1 \leq b$ by (12.18). Since $|a_0|_4 \leq q_1$ we have $\mathrm{enc}(q_1, |\mathrm{Enc}\, x|_4 \cdot q_1 \cdot 4 \cdot 4, x, a_1)$, so that $x \in a_1$. Also $a_0 \subseteq a_1$, so that $a \subseteq a_1$. Since a is a set we have $a \leq a_1$ by (12.9). Hence $a \leq b$, a contradiction. Thus (12.19). □

This result is not useful at present because we have no replacement principle to produce functions. Function symbols are much easier to come by than functions. Let Q_2'' be the current theory.

Metatheorem 12.1 Let T *be an extension of* Q_2'' *and let* f *be a bounded* $(1+\nu)$*-ary function symbol of* T. *Then the following is a theorem of* T:

$$\begin{aligned}\text{FS. } & a \text{ is a set } \ \& \\ & \forall x \forall y (x \in a \ \&\ y \in a \ \&\ x \neq y \rightarrow \mathrm{f}x x_1 \ldots x_\nu \neq \mathrm{f}y x_1 \ldots x_\nu)\ \& \\ & \forall x (x \in a \rightarrow x \leq \mathrm{f}x x_1 \ldots x_\nu \ \&\ \mathrm{f}x x_1 \ldots x_\nu \in b) \rightarrow a \leq b.\end{aligned}$$

Demonstration. To obtain a proof of (FS), replace, in the proof of (12.19), each occurrence of (12.19) by (FS) and each occurrence of $f(\mathrm{x})$ by $\mathrm{fx}x_1 \ldots x_\nu$.

12.20 *Def.* $\mathrm{Graph}(x, f) = \langle x, f(x) \rangle$.

12.21 *Thm.* a is a set & $\forall x(x \in a \rightarrow x$ is in the domain of $f) \rightarrow a \leq f$.

Proof. We have $x \neq y \rightarrow \mathrm{Graph}(x, f) \neq \mathrm{Graph}(y, f)$ by (12.11), and $x \leq \mathrm{Graph}(x, f)$ by (12.12), and x is in the domain of $f \rightarrow \mathrm{Graph}(x, f) \in f$. Suppose *hyp*(12.21). By FS we have $a \leq f$. Thus (12.21).

12.22 *Def.* $\mathrm{Dom}\, f = \mathrm{Max}\, a(a \leq f\ \&\ a$ is a set & $\forall x(x \in a \rightarrow x$ is in the domain of $f))$.

12.23 *Thm.* $x \in \mathrm{Dom}\, f \leftrightarrow x$ is in the domain of f.

Proof. Since $\mathit{scope}_{\mathrm{Max}\, a}(12.22)_a[0]$, we have *lhs* (12.23) $\rightarrow$ *rhs* (12.23) by MAX. Suppose *rhs* (12.23) & $\neg$*lhs* (12.23), and let $a = \mathrm{Dom}\, f \cup \{x\}$.

Then a is a set and $\forall x_1(x_1 \in a \rightarrow x_1$ is in the domain of $f)$, so $a \leq f$ by (12.21). By the maximality of $\operatorname{Dom} f$ we have $a \leq f$, which is impossible since $\operatorname{Dom} f \subseteq a$ & $\operatorname{Dom} f \neq a$. Thus (12.23).

12.24 *Def.* $\operatorname{Hparg}(y, f) = \langle \operatorname{Min} x(x \leq f \;\&\; f(x) = y), y\rangle$.

12.25 *Thm.* a is a set & $\forall y(y \in a \rightarrow y$ is in the range of $f) \rightarrow a \leq f$,

Proof. We have $y_1 \neq y_2 \rightarrow \operatorname{Hparg}(y_1, f) \neq \operatorname{Hparg}(y_2, f)$, and $y \leq \operatorname{Hparg}(y, f)$. Also, y is in the range of $f \rightarrow \operatorname{Hparg}(y, f) \in f$ by MIN. Suppose *hyp*(12.25). By FS we have $a \leq f$. Thus (12.25).

12.26 *Def.* $\operatorname{Ran} f = \operatorname{Max} a(a \leq f$ & a is a set & $\forall y(y \in a \rightarrow y$ is in the range of $f))$.

12.27 *Thm.* $y \in \operatorname{Ran} f \leftrightarrow y$ is in the range of f.

Proof. The proof is almost identical with that of (12.23). □

The notions of set and function are very useful, but we must take care not to use properties that have not been established. In particular, we do not have the set of all numbers $\leq n$, and we do not have power sets.

The intuition that the set of all subsets of a finite set is finite—or more generally, that if A and B are finite sets, then so is the set B^A of all functions from A to B—is a questionable intuition. Let A be the set of some 5000 spaces for symbols on a blank sheet of typewriter paper, and let B be the set of some 80 symbols of a typewriter; then perhaps B^A is infinite. Perhaps it is even incorrect to think of B^A as being a set. To do so is to postulate an entity, the set of all possible typewritten pages, and then to ascribe some kind of reality to this entity—for example, by asserting that one can in principle survey each possible typewritten page. But perhaps it simply is not so. Perhaps there is no such number as 80^{5000}; perhaps it is always possible to write a new and different page. Many ordinary activities are built up in a similar way from a rather small set of symbols or actions. Perhaps infinity is not far off in space or time or thought; perhaps it is while engaged in an ordinary activity—writing a page, getting a child ready for school, talking with someone, teaching a class, making love—that we are immersed in infinity.

Chapter 13

Exponential functions

In this chapter we prove some familiar properties of exponentiation by assuming as given a function f that is exponentiation ($i \mapsto x^i$) on the domain of all $i \leq k$.

13.1 *Def.* $\exp(x,k,f) \leftrightarrow f$ is a function & $f(0) = 1$ & $\forall i(i \in \operatorname{Dom} f \leftrightarrow i \leq k)$ & $\forall i(i < k \rightarrow f(i+1) = x \cdot f(i))$.

We have *rhs* (13.1): $i \leq \operatorname{Max}(f,k)$.

13.2 *Thm.* $\exp(x,k,f)$ & $\exp(x,k,g) \rightarrow f = g$.

Proof. By (12.17).

13.3 *Thm.* $\exp(x,k,f)$ & $\exp(y,k,g)$ & $\exp(x \cdot y,k,h) \rightarrow h(k) = f(k) \cdot g(k)$.

Proof. Suppose *hyp*(13.3). We will show that in fact 1: $i \leq k \rightarrow h(i) = f(i) \cdot g(i)$. Suppose $\exists i \neg(1)$. By BLNP there exists a minimal such i. Clearly $i \neq 0$. Therefore $h(i-1) = f(i-1) \cdot g(i-1)$, and so $h(i) = f(i) \cdot g(i)$, a contradiction. Thus (1), and thus (13.3).

13.4 *Thm.* $\exp(x,k,f)$ & $\exp(x,l,g)$ & $\exp(x,k+l,h) \rightarrow h(k+l) = f(k) \cdot g(l)$.

Proof. Suppose *hyp*(13.4). We will show that 1: $i \leq k \rightarrow h(i+l) = f(i) \cdot g(l)$. Suppose $\exists i \neg(1)$. By BLNP there exists a minimal such i. Clearly $i \neq 0$. Therefore $h(i-1+l) = f(i-1) \cdot g(l)$, and so $h(i+l) = f(i) \cdot g(l)$, a contradiction. Thus (1), and thus (13.4).

13.5 *Def.* $\mathrm{Rstr}(f,i) = \{w \in f : \exists j \exists y(\langle j,y\rangle = w \;\&\; j \le i)\}$.

13.6 *Thm.* f is a function $\to$ $\mathrm{Rstr}(f,i)$ is a function $\&$ $(j \le i \to \mathrm{Rstr}(f,i)(j) = f(j))$.

Proof. By BS.

13.7 *Thm.* $\exp(x,k,f) \;\&\; i \le k \to \exp(x,i,\mathrm{Rstr}(f,i))$.

Proof. By BS.

13.8 *Thm.* $\exp(x,k,f) \;\&\; \exp(f(k),l,g) \;\&\; \exp(x,k\cdot l,h) \to h(k\cdot l) = g(l)$.

Proof. Suppose hyp(13.8). We will show that 1: $i \le l \to h(k\cdot i) = g(i)$. Suppose $\exists i \neg(1)$. By BLNP there exists a minimal such i. Clearly $i \ne 0$. Therefore $h(k\cdot(i-1)) = g(i-1)$. By (13.7) and (13.4),

$$\mathrm{Rstr}(h,k\cdot i)(k\cdot i) = \mathrm{Rstr}(h,k\cdot(i-1))(k\cdot(i-1))\cdot\mathrm{Rstr}(h,k)(k),$$

so by (13.6), $h(k\cdot i) = h(k\cdot(i-1))\cdot h(k)$. That is, $h(k\cdot i) = g(i-1)\cdot h(k)$. But by (13.7) and (13.2), $\mathrm{Rstr}(h,k) = f$, so by (13.6), $h(k) = f(k)$. Therefore $h(k\cdot i) = g(i-1)\cdot f(k)$. But $g(i-1)\cdot f(k) = g(i)$, so $h(k\cdot i) = g(i)$, a contradiction. Thus (1), and thus (13.8).

13.9 *Thm.* $x \ne 0 \;\&\; \exp(x,k,f) \;\&\; i \le k \to f(i) \ne 0$.

Proof. Suppose $\exists i \neg$(13.9). By BLNP there exists a minimal such i. Clearly $i \ne 0$, so $f(i-1) \ne 0$ and $f(i) \ne 0$, a contradiction. Thus (13.9).

13.10 *Thm.* $2 \le x \;\&\; \exp(x,k,f) \;\&\; i < j \le k \to f(i) < f(j)$.

Proof. Suppose $\exists j \neg$(13.10). By BLNP there exists a minimal such j. Clearly $j \ne 0$, so $f(i) < f(j-1) \vee i = j-1$, and hence $f(i) \le f(j-1)$. But $f(j-1) \ne 0$ by (13.9), so $f(i) \le f(j-1) < x\cdot f(j-1) = f(j)$, a contradiction. Thus (13.10).

13.11 *Thm.* $2 \le x \;\&\; \exp(x,k_1,f_1) \;\&\; \exp(x,k_2,f_2) \;\&\; f_1(k_1) = f_2(k_2) \to k_1 = k_2 \;\&\; f_1 = f_2$.

Proof. Suppose hyp(13.11) and suppose $k_1 < k_2$. By (13.7) we have $\exp(x,k_1,\mathrm{Rstr}(f_2,k_1))$, so by (13.2), $\mathrm{Rstr}(f_2,k_1) = f_1$. By (13.6), $f_2(k_1) = f_1(k_1)$, but by (13.10), $f_2(k_1) < f_2(k_2)$. Hence $f_1(k_1) < f_2(k_2)$, a contradiction, and thus $k_2 \le k_1$. Similarly, $k_1 \le k_2$, so that $k_1 = k_2$. By (13.2), $f_1 = f_2$. Thus (13.11).

13.12 *Thm.* $\exp(2, k, f)$ & $i \leq k \to f(i)$ is a power of two.

Proof. Suppose $\exists i \neg (13.12)$. By BLNP there exists a minimal such i. Clearly $i \neq 0$, so $f(i-1)$ is a power of two and hence $f(i)$ is a power of two, a contradiction. Thus (13.12).

13.13 *Thm.* $\exp(2, k, f)$ & $x \leq f(k) \to \exists l (l \leq k$ & $|x|_2 = f(l))$.

Proof. Suppose *hyp*(13.13) and suppose $x \leq 1$. Then $|x|_2 = 1 = f(0)$, and thus $x \leq 1 \to$ *con* (13.13). Suppose $2 \leq x$. Then $|x|_2 \leq x$, so $|x|_2 \leq f(k)$. By BLNP there exists l such that $\min_l |x|_2 \leq f(l)$. But $|x|_2 \neq 1$, so $l \neq 0$ and hence $f(l-1) < |x|_2 \leq f(l) = 2 \cdot f(l-1)$. By (13.12), $f(l-1)$ and $|x|_2$ are powers of two, so by (10.6) we obtain $1 < |x|_2 / f(l-1) \leq 2$. Hence $|x|_2 = f(l)$. Thus *con* (13.13), and thus (13.13).

13.14 *Thm.* $\exp(x, k, f)$ & $\exp(y, k, g)$ & $x \leq y \to f(k) \leq g(k)$.

Proof. Suppose $\exists k \exists f \exists g \neg (13.14)$. By BLNP there exist minimal such k, f, and g. Clearly $k \neq 0$. Let $f_1 = \mathrm{Rstr}(f, k-1)$ and let $g_1 = \mathrm{Rstr}(g, k-1)$. By the minimality assumption, $f_1(k-1) \leq g_1(k-1)$. Therefore $f(k) \leq g(k)$, a contradiction. Thus (13.14).

Chapter 14

Exponentiation

Let Q_2''' be the current theory. Notice: in this chapter we will work in unbounded extensions of Q_2'''. (When I say that something is unbounded, I mean merely that I do not claim that it is bounded.) We mark with "!" the defining axiom of any unbounded symbol. The "!" serves as a warning that we may not use BI (or BLNP, BSD, BS, MIND, MIN, MAXD, MAX, or FS) on formulas containing the symbol.

14.1 *Def!* $\varepsilon(k) \leftrightarrow \forall x \exists f \exp(x, k, f)$.

14.2 *Thm.* $ind_k\, \varepsilon(k)$.

Proof. We have $\forall x \exp(x, 0, \{\langle 0, 1\rangle\})$ and so $\varepsilon(0)$. Suppose $\exp(x, k, f)$ and let $g = f \cup \{\langle \mathrm{S}k, x \cdot f(k)\rangle\}$. Then $\exp(x, \mathrm{S}k, g)$, and thus (14.2). □

Recall from Chapter 5 the abbreviations C^1, C^2, and C^3, where C is a unary formula. If p is a unary predicate, we write $\mathrm{p}^1(\mathrm{a})$ for $\mathrm{p}(k)^1[\mathrm{a}]$, $\mathrm{p}^2(\mathrm{a})$ for $\mathrm{p}(k)^2[\mathrm{a}]$, and $\mathrm{p}^3(\mathrm{a})$ for $\mathrm{p}(k)^3[\mathrm{a}]$.

14.3 *Thm.* $(\varepsilon^3(k) \rightarrow \varepsilon(k))$ & $(\varepsilon^3(k)$ & $i \leq k \rightarrow \varepsilon^3(i))$ & $\varepsilon^3(0)$ & $(\varepsilon^3(k) \rightarrow \varepsilon^3(\mathrm{S}k))$ & $(\varepsilon^3(k)$ & $\varepsilon^3(l) \rightarrow \varepsilon^3(k + l)$ & $\varepsilon^3(k \cdot l))$.

Proof. By (14.2) and REL.

14.4 *Def!* $x \uparrow k = z \leftrightarrow \varepsilon^3(k)$ & $\exists f(\exp(x, k, f)$ & $f(k) = z)$, otherwise $z = 0$.

The uniqueness condition holds by (13.2). It is convenient to have a symbol $\uparrow$ for exponentiation, but we also use the abbreviation a^{b} for $\mathrm{a} \uparrow \mathrm{b}$.

Let us use semantics to discuss the picture of the number system that is emerging. The discussion will be informal; we will not distinguish between a model and its universe, or between individuals and their names (see [Sh,§2.5]). Let T be the theory whose nonlogical axioms are (3.1)–(14.4) together with all formulas of the form (BSD), (MIND), and (MAXD) tacitly used in these formulas (but not the axiom scheme (MBI)), and whose nonlogical symbols are those occurring in these formulas. Let M be a model of T. For u a nonlogical symbol of T, let us also denote the corresponding predicate or function of M by u. Let C be an inductive formula of T and let $\mathrm{M_C} = \{\xi \in \mathrm{M} : \mathrm{C}^3[\xi]\}$. This becomes a model of T if we make the following definitions. Each nonlogical predicate symbol p of T is introduced in an axiom of the form $\mathrm{px_1 \ldots x_\nu} \leftrightarrow \mathrm{D}$; for $\xi_1, \ldots, \xi_\nu$ in $\mathrm{M_C}$, we say that $\mathrm{p_C}(\xi_1, \ldots, \xi_\nu)$ holds in case $\mathrm{D}^{\mathrm{C}^3}[\xi_1, \ldots, \xi_\nu]$ holds. The set $\mathrm{M_C}$ contains 0 and is closed under S, P, +, and $\cdot$. Every other function symbol f of T is introduced in an axiom of the form $\mathrm{fx_1 \ldots x_\nu = y} \leftrightarrow \mathrm{D}$; for $\xi_1, \ldots, \xi_\nu$ in $\mathrm{M_C}$ we let $\mathrm{f_C}(\xi_1, \ldots, \xi_\nu)$ be that element η of $\mathrm{M_C}$ such that $\mathrm{D}^{\mathrm{C}^3}[\xi_1, \ldots, \xi_\nu, \eta]$ holds. Now for p a bounded predicate symbol (i.e., for p a predicate symbol other than ε), $\mathrm{p_C}$ is simply the restriction of p to $\mathrm{M_C}$, and for f a bounded function symbol (i.e., for f a function symbol other than $\uparrow$), $\mathrm{f_C}$ is simply the restriction of f to $\mathrm{M_C}$. A bounded predicate symbol does not change its meaning; a bounded function symbol does not change its values. It is different for ε and $\uparrow$. Suppose we have an individual κ in $\mathrm{M_C}$ such that $\varepsilon(\kappa)$ holds. Then for all ξ in M, and so certainly for all ξ in $\mathrm{M_C}$, there is a ϕ such that $\exp(\xi, \kappa, \phi)$ holds—but nothing guarantees that $\phi \in \mathrm{M_C}$ even when ξ and κ are in $\mathrm{M_C}$. But $\varepsilon_\mathrm{C}(\kappa)$ holds only if $\phi \in \mathrm{M_C}$, so that $\varepsilon(\kappa)$ and $\varepsilon_\mathrm{C}(\kappa)$ may mean different things. Similarly, for ξ and κ in $\mathrm{M_C}$ we may have $\xi \uparrow \kappa = \varsigma$ with $\varsigma \notin \mathrm{M_C}$. In this case $\xi \uparrow_\mathrm{C} \kappa = 0$, so that $\xi \uparrow \kappa$ and $\xi \uparrow_\mathrm{C} \kappa$ may have different values.

Now if the inductive formula C is bounded, then C holds in the model $\mathrm{M_C}$; this is the method we have been using to successively refine our concept of number. Each time we perform a bounded induction the model becomes smaller, and closer to our intuitive idea of what the number system should be. But for an unbounded C this need not be so. For example, consider the inductive formula $\varepsilon(\kappa)$, which we call ε for short, and form M_ε. We have $\varepsilon(\kappa)$ for all κ in M_ε, but this does not mean that we have a model of $\mathrm{T}[\forall k\, \varepsilon(k)]$. We can iterate this process: introduce $\varepsilon_1, \varepsilon_2, \ldots$ by

$$\varepsilon_1(k) \leftrightarrow \forall x(\varepsilon^3(x) \to \exists f(\varepsilon^3(f) \mathbin{\&} \exp(x,k,f))),$$

$$\varepsilon_2(k) \leftrightarrow \forall x(\varepsilon_1^3(x) \to \exists f(\varepsilon_1^3(f) \mathbin{\&} \exp(x,k,f))),$$

$$\ldots.$$

In this way we obtain successively smaller models,

$$\mathrm{M} \supseteq \mathrm{M}_\varepsilon \supseteq \mathrm{M}_{\varepsilon_1} \supseteq \cdots \supseteq \mathrm{M}_{\varepsilon_\nu} \supseteq \cdots,$$

but none of these relativizations helps one whit to establish $\forall k\, \varepsilon(k)$.

Semantics tells us that there is an intersection $\mathrm{M}_{\varepsilon_\omega}$ of these models and that $\mathrm{M}_{\varepsilon_\omega}$ is closed under exponentiation. This seems to be beyond the reach of syntax, as is the minimal model N. The minimal model is a model of I, since for any inductive C we have $\mathrm{N_C} \subseteq \mathrm{N}$ and so $\mathrm{N_C} = \mathrm{N}$. This N is the legendary standard model of Peano Arithmetic. But how can we express what membership in N means? The concept of the finite in mathematics is subtle and elusive.

Semantics is a picturesque metaphor to illustrate syntax, but like all metaphors it depicts more than may have been intended. A nominalist distrusts semantical reasoning that involves syntactically inexpressible concepts.

14.5 *Thm.* $\varepsilon^3(k) \mathbin{\&} \varepsilon^3(l) \to (x \cdot y)^k = x^k \cdot y^k \mathbin{\&} x^{k+l} = x^k \cdot x^l \mathbin{\&} x^{k\,l} = (x^k)^l$.

Proof. By (14.3), (13.3), (13.4), and (13.8).

14.6 *Def!* x is a power of $b \leftrightarrow \exists k(\varepsilon^3(k) \mathbin{\&} b^k = x)$.

14.7 *Thm.* x is a power of $2 \to x$ is a power of two.

Proof. Suppose x is a power of 2. Then there exist k and f such that $\exp(2,k,f) \mathbin{\&} f(k) = x$. By (13.12), x is a power of two. Thus (14.7). □

What about the converse to (14.7)—can we show that 1: x is a power of two $\to x$ is a power of 2? If x is a power of two, we want a k such that $2^k = x$. This k will be much smaller than x, so this does not seem too much to ask. But we cannot use BLNP to prove (1) because $2^k = x$ is an unbounded formula. This suggests, not that there is an arithmetical obstacle, but that our formalism is inadequate. The problem of establishing (1) is a test problem; the motivation for attacking it is to obtain more powerful methods. The difficulty is that the objects used to express the relationship

$2^k = x$, in particular the function f such that $\exp(2, k, f)$ & $f(k) = x$, are big. We can obtain a bound on f of the form $2^{C \cdot k \cdot k}$, where C is a constant, but this cannot be polynomially bounded in terms of x (that is, 2^k). Let us try to go beyond polynomials and use the function # that on powers of 2 satisfies $2^k \# 2^l = 2^{k \cdot l}$. Then powers of 2 behave with respect to 1, multiplication by 2, $\cdot$, and # like numbers with respect to 0, S, +, and $\cdot$, so we should be able to adapt the relativization scheme of Chapter 5 to include #.

14.8 *Def!* $\lambda(x) \leftrightarrow \exists k\ x \leq 2^k$.

We would hope to establish that $\lambda(x)$ holds universally, but let us first investigate the universe of numbers x that do satisfy $\lambda(x)$.

14.9 *Thm.* $\lambda(0)$ & $(\lambda(x) \rightarrow \lambda(\mathrm{S}x)$ & $\lambda(\mathrm{P}x))$ & $(\lambda(x)$ & $\lambda(y) \rightarrow \lambda(x+y)$ & $\lambda(x \cdot y))$ & $(\lambda(x)$ & $w \leq x \rightarrow \lambda(w))$.

Proof. Clearly $\lambda(0)$. Suppose $x \leq 2^k$ & $y \leq 2^l$ & $w \leq x$. Then $\mathrm{P}x \leq \mathrm{S}x \leq 2^{k+1}$, so $\lambda(\mathrm{P}x)$ and $\lambda(\mathrm{S}x)$. We have $x + y \leq 2 \uparrow (\mathrm{Max}(k,l)+1)$, so $\lambda(x+y)$. We have $x \cdot y \leq 2^{k+l}$, so $\lambda(x \cdot y)$. Finally, $\lambda(w)$ is obvious. Thus (14.9).

14.10 *Def!* $\mathrm{Log}_0\, x = k \leftrightarrow |x|_2 = 2^k$, otherwise $k = 0$.

The uniqueness condition holds by (13.11).

14.11 *Thm.* $\varepsilon^3(\mathrm{Log}_0\, x)$.

Proof. We have $\mathrm{Log}_0\, x = 0 \rightarrow$ (14.11) since $\varepsilon^3(0)$, so suppose $\mathrm{Log}_0\, x \neq 0$. Since $|x|_2 \neq 0$ we have $\varepsilon^3(\mathrm{Log}_0\, x)$ from (14.4), and thus (14.11).

14.12 *Thm.* $\lambda(2^k)$.

Proof. From (14.8).

14.13 *Thm.* $\lambda(x) \rightarrow |x|_2 = 2^{\mathrm{Log}_0\, x} = 2^{\mathrm{Log}_0 |x|_2}$.

Proof. Suppose $\lambda(x)$. Then there exist k and f such that $\varepsilon^3(k)$ & $\exp(2, k, f)$ & $x \leq f(k)$. By (13.13) there exists l such that $l \leq k$ & $|x|_2 = f(l)$. We have $\varepsilon^3(l)$. Also, $\exp(2, l, \mathrm{Rstr}(f, l))$ and $\mathrm{Rstr}(f,l)(l) = f(l) = |x|_2$, so that $|x|_2 = 2^l$ and $l = \mathrm{Log}_0\, x = \mathrm{Log}_0 |x|_2$. Thus (14.13).

14.14 *Def!* $x \#_0 y = 2^{\mathrm{Log}_0\, x \cdot \mathrm{Log}_0\, y}$.

14.15 *Thm.* $\lambda(x \#_0 y)$.

Proof. By (14.12).

14.16 *Thm.* $x \#_0 y = |x \#_0 y|_2 = |x|_2 \#_0 |y|_2 = |x|_2 \#_0 y = x \#_0 |y|_2$.

Proof. We have

$$x \#_0 y = 2 \uparrow (\mathrm{Log}_0\, x \cdot \mathrm{Log}_0\, y) =$$
$$2 \uparrow (\mathrm{Log}_0\, |x|_2 \cdot \mathrm{Log}_0\, |y|_2) = |x|_2 \#_0 |y|_2.$$

By (14.11) and (14.3) we have $\varepsilon^3(\mathrm{Log}_0\, x \cdot \mathrm{Log}_0\, y)$, so by (14.7), $x \#_0 y$ is a power of two. Hence $x \#_0 y = |x \#_0 y|_2$. Since $||x|_2|_2 = |x|_2$ we have (14.16).

14.17 *Thm.* $x \#_0 1 = 1$.

Proof. We have $x \#_0 1 = 2 \uparrow (\mathrm{Log}_0\, x \cdot \mathrm{Log}_0\, 1) = 2 \uparrow 0 = 1$.

14.18 *Thm.* $\lambda(x) \rightarrow x \#_0 2 = |x|_2$.

Proof. We have $x \#_0 2 = 2 \uparrow (\mathrm{Log}_0\, x \cdot \mathrm{Log}_0\, 2) = 2 \uparrow \mathrm{Log}_0\, x$, so by (14.13) we have (14.18).

14.19 *Thm.* $x \#_0 y = y \#_0 x$.

Proof. We have $x \#_0 y = 2 \uparrow (\mathrm{Log}_0\, x \cdot \mathrm{Log}_0\, y) = 2 \uparrow (\mathrm{Log}_0\, y \cdot \mathrm{Log}_0\, x) = y \#_0 x$.

14.20 *Thm.* $\mathrm{Log}_0\, (x \#_0 y) = \mathrm{Log}_0\, x \cdot \mathrm{Log}_0\, y$.

Proof. By (14.16), $|x \#_0 y|_2 = x \#_0 y = 2 \uparrow (\mathrm{Log}_0\, x \cdot \mathrm{Log}_0\, y)$, so (14.20).

14.21 *Thm.* $(x \#_0 y) \#_0 z = x \#_0 (y \#_0 z)$.

Proof. By (14.20) we get $(x \#_0 y) \#_0 z = 2 \uparrow (\mathrm{Log}_0\, (x \#_0 y) \cdot \mathrm{Log}_0\, z) = 2 \uparrow (\mathrm{Log}_0\, x \cdot \mathrm{Log}_0\, y \cdot \mathrm{Log}_0\, z) = 2 \uparrow (\mathrm{Log}_0\, x \cdot \mathrm{Log}_0\, (y \#_0 z)) = x \#_0 (y \#_0 z)$.

14.22 *Thm.* $\lambda(y)\ \&\ \lambda(z) \rightarrow x \#_0 (|y|_2 \cdot |z|_2) = (x \#_0 |y|_2) \cdot (x \#_0 |z|_2)$.

Proof. Suppose $\lambda(y)\ \&\ \lambda(z)$. By (14.13), $|y|_2 = 2 \uparrow \mathrm{Log}_0\, |y|_2$ and $|z|_2 = 2 \uparrow \mathrm{Log}_0\, |z|_2$, so that $|y|_2 \cdot |z|_2 = 2 \uparrow (\mathrm{Log}_0\, |y|_2 + \mathrm{Log}_0\, |z|_2)$ and so $\mathrm{Log}_0\, (|y|_2 \cdot |z|_2) = \mathrm{Log}_0\, |y|_2 + \mathrm{Log}_0\, |z|_2$. Hence

$$x \#_0 (|y|_2 \cdot |z|_2) = 2 \uparrow (\mathrm{Log}_0\, x \cdot (\mathrm{Log}_0\, |y|_2 + \mathrm{Log}_0\, |z|_2)) =$$
$$(2 \uparrow (\mathrm{Log}_0\, x \cdot \mathrm{Log}_0\, |y|_2)) \cdot (2 \uparrow (\mathrm{Log}_0\, x \cdot \mathrm{Log}_0\, |z|_2)) =$$
$$(x \#_0 |y|_2) \cdot (x \#_0 |z|_2).$$

Thus (14.22).

14.23 *Thm.* $\lambda(z) \rightarrow (y \leq z \rightarrow x \#_0 y \leq x \#_0 z)$.

Proof. Suppose $\lambda(z)$ & $y \leq z$. Then $\lambda(y)$, and by (14.13) and (13.10) we have $\mathrm{Log}_0\, y \leq \mathrm{Log}_0\, z$. By (13.10) again, $x \#_0 y = 2 \uparrow (\mathrm{Log}_0\, x \cdot \mathrm{Log}_0\, y) \leq 2 \uparrow (\mathrm{Log}_0\, x \cdot \mathrm{Log}_0\, z) = x \#_0 z$. Thus (14.23).

Chapter 15

A stronger relativization scheme

We concluded Chapter 14 by proving some properties of $\#_0$, but some of these held only conditionally, subject to $\lambda(x)$ for certain x. What we can't prove, we can postulate. Let Q_3 be the theory obtained from Q_2''' (this is the bounded extension of Q_2 defined at the beginning of Chapter 14) by adjoining a binary function symbol $\#$ and the following nonlogical axioms:

15.1 *Ax.* $x \# y = |x \# y|_2 = |x|_2 \# |y|_2 = |x|_2 \# y = x \# |y|_2$,

15.2 *Ax.* $x \# 1 = 1$,

15.3 *Ax.* $x \# 2 = 2$,

15.4 *Ax.* $x \# y = y \# x$,

15.5 *Ax.* $(x \# y) \# z = x \# (y \# z)$,

15.6 *Ax.* $x \# (|y|_2 \cdot |z|_2) = (x \# |y|_2) \cdot (x \# |z|_2)$,

15.7 *Ax.* $y \leq z \ \rightarrow x \# y \leq x \# z$.

Metatheorem 15.1 Q_3 is interpretable in Q_2.

Demonstration. Let T be the (unbounded) extension by definitions of Q_2 obtained by adjoining all of the symbols and defining axioms through the end of Chapter 14. We construct an interpretation I of Q_3 in T. Let the universe of I be λ. For each nonlogical symbol u of Q_3 other than $\#$, let u_I

be u, and let $\#_1$ be $\#_0$. By (14.9), λ is hereditary and respects 0, S, P, +, and $\cdot$. Also, λ respects each nonlogical axiom of Q_2: the nonlogical axioms of Q_1 because they are open, the defining axiom of $\leq$ because λ is hereditary, and the axioms (MBI) by Metatheorem 7.1. By Metatheorem 7.7, λ respects each nonlogical axiom and function symbol of Q_2'''. By (14.15) we have $\vdash_T \lambda(x)\ \&\ \lambda(y) \rightarrow \lambda(x \#_1 y)$. The interpretations of the axioms (15.1)–(15.7) are theorems of T by (14.16)–(14.19) and (14.21)–(14.23). □

If C is a unary formula, we write

$$C^4 \text{ for } \forall y(C^3[y] \rightarrow C^3[y \# x]).$$

Metatheorem 15.2 Let T *be an extension of* Q_3 *and let* C *be a unary formula of* T. *Then the following is a theorem of* T:

$$\begin{aligned} \text{SREL.}\quad & C[0]\ \&\ \forall x(C[x] \rightarrow C[Sx]) \rightarrow \\ & (C^4[x] \rightarrow C[x])\ \& \\ & (C^4[x]\ \&\ w \leq x \rightarrow C^4[w])\ \& \\ & C^4[0]\ \& \\ & (C^4[x] \rightarrow C^4[Sx]\ \&\ C^4[Px])\ \& \\ & (C^4[x_1]\ \&\ C^4[x_2] \rightarrow C^4[x_1 + x_2]\ \&\ C^4[x_1 \cdot x_2]\ \&\ C^4[x_1 \# x_2]). \end{aligned}$$

Demonstration. We prove (SREL) in T as follows. Suppose *hyp*(SREL). Suppose $C^4[x]$. We have $C^3[2]$ by REL, and consequently $C^3[2 \# x]$. By (15.4) and (15.3), $2 \# x = |x|_2$, and so $C^3[|x|_2]$. But $x \leq |x|_2 \cdot x$, so by REL, $C^3[x]$ and hence $C[x]$. Thus $C^4[x] \rightarrow C[x]$. Suppose $C^4[x]\ \&\ w \leq x\ \&\ C^3[y]$. Then $C^3[y \# x]$. But $y \# w \leq y \# x$ by (15.7), so $C^3[y \# w]$ by REL. Thus $C^4[x]\ \&\ w \leq x \rightarrow C^4[w]$. By (15.1) and (15.2), $y \# 0 = y \# |0|_2 = y \# 1 = 1$, and $C^1[1]$ by REL. Hence $C^4[0]$. Let $a = |x_1|_2 \cdot 2 \cdot |x_2|_2 \cdot 2$. Since $Sx_1 \leq |x_1|_2 \cdot 2$ we have $Sx_1 \leq a$, $Px_1 \leq a$, and $x_1 \cdot x_2 \leq a$; and since $x_1 + x_2 \leq x_1 \cdot 2 \vee x_1 + x_2 \leq x_2 \cdot 2$ we also have $x_1 + x_2 \leq a$. Suppose $C^4[x_1]\ \&\ C^4[x_2]\ \&\ C^3[y]$. Now

$$\begin{aligned} & y \# a = (y \# |x_1|_2) \cdot (y \# 2) \cdot (y \# |x_2|_2) \cdot (y \# 2) = \\ & (y \# x_1) \cdot |y|_2 \cdot (y \# x_2) \cdot |y|_2 \leq (y \# x_1) \cdot SPy \cdot (y \# x_2) \cdot SPy \end{aligned}$$

by (15.6), (15.1), and (15.2). But we have $C^3[y \# x_1]$ and $C^3[y \# x_2]$, so by REL we $C^3[y \# a]$. By (15.7) and REL we have $C^3[y \# Sx_1]$, $C^3[y \# Px_1]$, $C^3[y \# (x_1 + x_2)]$, and $C^3[y \# (x_1 \cdot x_2)]$. Finally, we have $C^3[y \# x_1]$ and hence $C^3[(y \# x_1) \# x_2]$, so we have $C^3[y \# (x_1 \# x_2)]$ by (15.5). Thus *con*(SREL), and thus (SREL). □

This relativization scheme is also due to Solovay; see [PD] and [Pu].

Metatheorem 15.3 Let A *be a manifestly bounded formula of* Q_3 *that is inductive in one of its free variables* x. *Then* $Q_3[\mathrm{A}]$ *is interpretable in* Q_3.

Demonstration. Let C be $\mathrm{A}_{/\mathrm{x}}$. By SREL, C^4 is hereditary and respects 0, S, P, +, and ·. Also, C^4 respects each nonlogical axiom of Q_2: the nonlogical axioms of Q_1 because they are open, the defining axiom of $\leq$ because C^4 is hereditary, and the axioms (MBI) by Metatheorem 7.1. By Metatheorem 7.7, C^4 respects each nonlogical axiom and function symbol of Q_2'''. By SREL, C^4 respects #, and it respects (15.1)–(15.7) because they are open. Finally, C^4 respects A by Metatheorem 7.1. Therefore $Q_3[\mathrm{A}]$ is interpretable in Q_3 via the interpretation associated with C^4. □

We construct a theory Q_4, which is an extension of Q_3 with the same language, by adjoining as new nonlogical axioms all formulas of the form (MBI) (see Chapter 7) where now A is a manifestly bounded formula in the language of Q_3.

Metatheorem 15.4 Let $\mathrm{B}_1, \ldots, \mathrm{B}_\lambda$ *be theorems of* Q_4. *Then the theory* $Q_3[\mathrm{B}_1, \ldots, \mathrm{B}_\lambda]$ *is interpretable in* Q_3.

Demonstration. In the demonstration of Metatheorem 7.3, replace each occurrence of Q_2 by Q_4, each occurrence of Q_1' by Q_3, and the reference to Metatheorem 7.2 by a reference to Metatheorem 15.3.

Metatheorem 15.5 Q_4 *is a bounded theory.*

Demonstration. In the demonstration of Metatheorem 7.4, replace each occurrence of Q_2 by Q_4.

Metatheorem 15.6 Let U *be an extension of* Q_4 *and let* A *be a formula of* U *that is bounded over* Q_4. *Then*

BI. $ind_{\mathrm{x}}\,\mathrm{A} \rightarrow \mathrm{A}$

is a theorem of U.

Demonstration. By Metatheorems 15.5 and 7.5. □

Until possible further notice, when we say that a formula, nonlogical symbol, or term is bounded, we mean that it is bounded over Q_4. We have BLNP (Chap. 8), BSD, BS, MIND, MIN, MAXD, MAX (Chap. 11), and FS (Chap. 12) for bounded formulas. We will work in extensions by definition of Q_4, marking with "!" each defining axiom of an unbounded symbol. To start with, we adjoin the symbols and defining axioms of Chapter 14.

Putting all of our interpretability results together, and using the fact that an extension by definitions of a theory containing a constant is interpretable in it (see [Sh,§4.7]), we see that any extension by definitions of Q_4 is locally interpretable in Q.

Why could we not have settled at the beginning on a theory, called Predicative Arithmetic, with all of the necessary axioms? We would like to have a formula A in the language of Q be a theorem of Predicative Arithmetic if and only if $Q[\mathrm{A}]$ is interpretable in Q. Perhaps this is possible, but I do not know the answer to the following *compatibility problem*: if $Q[\mathrm{A}]$ and $Q[\mathrm{B}]$ are interpretable in Q, then is $Q[\mathrm{A}, \mathrm{B}]$ interpretable in Q? In the absence of a positive solution to the compatibility problem, it is necessary to take care, as we have, that our methods for adding a formula to the growing list of predicatively established formulas do not interfere with any previously established formula.

Chapter 16

Bounds on exponential functions

Now we take up the question of how big f is when $\exp(x, k, f)$ holds.

16.1 *Thm.* $\{x\} \le 146 \cdot (\mathrm{SP}x)^2$.

Proof. From the defining axioms of the function symbols involved, we find that $|x|_2 \le \mathrm{SP}x$ and $\mathrm{Enc}\, x \le \mathrm{SP}x \cdot \mathrm{SP}x \cdot 4$, and so we have $\{x\} \le 2 \cdot (\mathrm{SP}x \cdot \mathrm{SP}x \cdot 4) \cdot 4 \cdot 4 + (\mathrm{SP}x \cdot \mathrm{SP}x \cdot 4) \cdot 4 + 2 \le 146 \cdot (\mathrm{SP}x)^2$.

16.2 *Thm.* $a \cup b \le 5 \cdot \mathrm{SP}a \cdot \mathrm{SP}b$.

Proof. We have $a \cup b \le a \cdot \mathrm{SP}b \cdot 4 + b \le 5 \cdot \mathrm{SP}a \cdot \mathrm{SP}b$.

16.3 *Thm.* $\langle x, y \rangle \le 5 \cdot (\mathrm{Max}(x, y))^2$.

Proof. We have $\langle x, y \rangle = (x + y)^2 + y \le 4 \cdot (\mathrm{Max}(x, y))^2 + y^2 \le 5 \cdot (\mathrm{Max}(x, y))^2$.

16.4 *Def.* $\mathrm{K} = 18250$.

16.5 *Thm.* $f \cup \{\langle x, y \rangle\} \le \mathrm{K} \cdot \mathrm{SP}f \cdot (\mathrm{SP}\,\mathrm{Max}(x, y))^4$.

Proof. By (16.1), (16.2), and (16.3).

16.6 *Thm.* $2 \le x \;\&\; \exp(x, k, f) \;\&\; i \le k \rightarrow i \le f(i)$.

Proof. Suppose $\exists i \neg(16.6)$. By BLNP there exists a minimal such i. Clearly $i \ne 0$. Therefore $i - 1 \le f(i-1)$, and $f(i-1) \ne 0$ by (13.9), so $i \le x \cdot f(i-1) = f(i)$, a contradiction. Thus (16.6).

16.7 *Thm.* $2 \leq x \;\&\; \exp(x,k,f) \rightarrow f \leq \mathrm{K} \cdot \mathrm{Rstr}(f,k-1) \cdot f(k)^4$.

Proof. Suppose *hyp* (16.7). Then $f = \mathrm{Rstr}(f,k-1) \cup \{\langle k, f(k)\rangle\}$, so *con* (16.7) by (16.5), (16.6), and (13.9). Thus (16.7).

16.8 *Thm.* $a \# (2 \cdot b) = |a|_2 \cdot (a \# b)$.

Proof. We have $a \# (2 \cdot b) = a \# |2 \cdot b|_2 = a \# (2 \cdot |b|_2) = (a \# 2) \cdot (a \# |b|_2) = |a|_2 \cdot (a \# b)$.

16.9 *Def.* $x \geq y \leftrightarrow y \leq x$.

16.10 *Thm.* $2 \leq z \rightarrow \mathrm{K} \cdot z^4 \leq \mathrm{K} \# (2 \cdot z)$.

Proof. Suppose $2 \leq z$. Observe that $2^9 = 512 \leq \mathrm{K}$. Then $\mathrm{K} \# (2 \cdot z) = |\mathrm{K}|_2 \cdot (\mathrm{K} \# z) \geq |\mathrm{K}|_2 \cdot (2^9 \# z) = |\mathrm{K}|_2 \cdot |z|_2^9 = |\mathrm{K}|_2 \cdot |z|_2 \cdot (|z|_2^2)^4 \geq \mathrm{K} \cdot z^4$. Thus (16.10).

16.11 *Thm.* $2 \leq x \;\&\; \exp(x,k,f) \rightarrow f \leq \mathrm{K} \# (2 \cdot f(k)) \# (2 \cdot f(k))$.

Proof. Suppose $\exists k \exists f \neg$(16.11). By BLNP there exist minimal such k and f. Suppose $k = 0$. Then $f = \{\langle 0,1\rangle\} = \{(0+1)^2+1\} = \{2\} \leq 146 \cdot 4 = 584$, but $\mathrm{K} \# 2 \# 2 = |\mathrm{K}|_2 = 16394$, so *con* (16.11), a contradiction. Thus $k \neq 0$ and therefore

$$\mathrm{Rstr}(f,k-1) \leq \mathrm{K} \# (2 \cdot f(k-1)) \# (2 \cdot f(k-1)) \leq \mathrm{K} \# f(k) \# f(k).$$

By (16.7) and (16.10), $f \leq (\mathrm{K} \# (2 \cdot f(k))) \cdot (\mathrm{K} \# f(k) \# f(k))$. But

$$\begin{aligned} \mathrm{K} \# (2 \cdot f(k)) \# (2 \cdot f(k)) &= |\mathrm{K} \# (2 \cdot f(k))|_2 \cdot (\mathrm{K} \# (2 \cdot f(k)) \# f(k)) \\ &= (\mathrm{K} \# (2 \cdot f(k))) \cdot (\mathrm{K} \# (2 \cdot f(k)) \# f(k)) \end{aligned}$$

by (16.8), so *con* (16.11), a contradiction. Thus (16.11).

16.12 *Thm.* x is a power of two $\rightarrow \exists k \exists f(\exp(2,k,f) \;\;\&\;\; f(k) = x)$.

Proof. We have (16.12): $k \leq x, f \leq \mathrm{K} \# (2 \cdot x) \# (2 \cdot x)$ by (16.6) and (16.11). Suppose $\exists x \neg$(16.12). By BLNP there exists a minimal such x. Since $\exp(2,0,\{\langle 0,1\rangle\})$ and $\{\langle 0,1\rangle\}(0) = 1$ we have $x \neq 1$. Therefore $x/2$ is a power of two, and there exist k_1 and f_1 such that $\exp(2,k_1,f_1) \;\&\; f_1(k_1) = x/2$. Let $k = k_1 + 1$ and let $f = f_1 \cup \{\langle k, x\rangle\}$. Then $\exp(2,k,f) \;\&\; f(k) = x$, a contradiction. Thus (16.12). □

This does not yet solve the test problem of Chapter 14 because we need $\varepsilon^3(k)$ in order to have $2^k = x$.

16.13 *Thm.* $\varepsilon(k) \leftrightarrow \varepsilon^1(k)$.

Proof. Clearly $\varepsilon^1(k) \to \varepsilon(k)$. Suppose $\exp(x, k, f)$ & $i \le k$. Then $\exp(x, i, \mathrm{Rstr}(f, i))$ and thus $\varepsilon(k) \to \varepsilon^1(k)$. □

We need some sort of bound on f when $\exp(0, k, f)$ or $\exp(1, k, f)$ holds.

16.14 *Thm.* $\exp(0, k, f)$ & $0 < i \le k \to f(i) = 0$.

Proof. Suppose *hyp* (16.14). Then $f(i) = 0 \cdot f(i-1) = 0$. Thus (16.14).

16.15 *Thm.* $\exp(1, k, f)$ & $i \le k \to f(i) = 1$.

Proof. Suppose $\exists i \neg$(16.15). By BLNP there exists a minimal such i. Clearly $1 \ne 0$. Therefore $f(i-1) = 1$, so $f(i) = 1$, a contradiction. Thus (16.15).

16.16 *Def.* $\mathrm{Proj}_1\, z = x \leftrightarrow \exists y\, \langle x, y\rangle = z$, otherwise $x = 0$.

16.17 *Def.* $\mathrm{Proj}_2\, z = y \leftrightarrow \exists x\, \langle x, y\rangle = z$, otherwise $y = 0$.

16.18 *Def.* $\mathrm{Rplc}(z, g) = \langle \mathrm{Proj}_1\, z, g(\mathrm{Proj}_1\, z)\rangle$.

16.19 *Thm.* $x_1 \le x_2$ & $y_1 \le y_2 \to \langle x_1, y_1\rangle \le \langle x_2, y_2\rangle$.

Proof. Suppose *hyp* (16.19). There exist u and v such that $x_1 + u = x_2$ & $y_1 + v = y_2$. Therefore $\langle x_2, y_2\rangle = (x_1 + u + y_1 + v)^2 + y_1 + v = \langle x_1, y_1\rangle + (u+v)^2 + 2 \cdot (u+v) \cdot (x_1 + y_1) + v$. Thus (16.19).

16.20 *Thm.* $x < 2$ & $\exp(x, k, f)$ & $\exp(2, k, g) \to f \le g$.

Proof. Suppose *hyp* (16.20). Then 1: $z_1 \in f$ & $z_2 \in f$ & $z_1 \ne z_2 \to \mathrm{Rplc}(z_1, g) \ne \mathrm{Rplc}(z_2, g)$, since *hyp* (1) $\to \mathrm{Proj}_1\, z_1 \ne \mathrm{Proj}_1\, z_2$. Also, $z \in f \to z \le \mathrm{Rplc}(z, g)$ & $\mathrm{Rplc}(z, g) \in g$ by (16.19). By FS (see Chapter 12) we have *con* (16.20). Thus (16.20).

16.21 *Thm.* $\exp(2, k, f) \to \exists f_0 \exp(0, k, f_0)$ & $\exists f_1 \exp(1, k, f_1)$.

Proof. Now (16.20) gives (16.21): $f_0 \le f, f_1 \le f$. Suppose $\exists k \exists f \neg$(16.21). By BLNP there exist minimal such k and f. Clearly $k \ne 0$. Therefore $\exp(2, k-1, \mathrm{Rstr}(f, k-1))$, and there exist g_0 and g_1 such that we have $\exp(0, k-1, g_0)$ & $\exp(1, k-1, g_1)$. So now we let $f_0 = g_0 \cup \{\langle k, 0\rangle\}$ and let $f_1 = g_1 \cup \{\langle k, 1\rangle\}$. Then *con* (16.21), a contradiction, and thus (16.21).

16.22 *Thm.* $2 \le x$ & $\exp(x, k, f)$ & $\exp(x, l, g) \to \exists h \exp(x, k+l, h)$.

Proof. Observe that (16.22): $h \le \mathrm{K} \# (2 \cdot f(k) \cdot g(k)) \# (2 \cdot f(k) \cdot g(k))$ by (16.11) and (13.4). Suppose $\exists k \exists f \neg$(16.22). By BLNP there exist minimal such k and f. Obviously $k \neq 0$. We have $\exp(x, k-1, \mathrm{Rstr}(f, k-1))$, so there exists h_1 such that $\exp(x, k-1+l, h_1)$. Let $h = h_1 \cup \{\langle k+l, f(k) \cdot g(l)\rangle\}$. Then $\exp(x, k+l, h)$, which is a contradiction. Thus (16.22).

16.23 *Thm.* $\varepsilon(k) \leftrightarrow \varepsilon^2(k)$.

Proof. Clearly $\varepsilon^2(k) \to \varepsilon(k)$ (or if it is not clear, see the proof of REL). Suppose $\varepsilon(k)$ & $\varepsilon(l)$. We claim that 1: $\exists h \exp(x, k+l, h)$. Suppose $x < 2$. There exist f and g such that $\exp(2, k, f)$ & $\exp(2, l, g)$, so by (16.22) there exists h_1 such that $\exp(2, k+l, h_1)$. By (16.21) we have (1), and thus $x < 2 \to (1)$. Suppose $2 \leq x$. There exist f and g such that $\exp(x, k, f)$ & $\exp(x, l, g)$, so by (16.22) we have (1). Thus (1), and thus $\varepsilon(k)$ & $\varepsilon(l)$ $\to \varepsilon(k+l)$. By (16.13) we have $\varepsilon(k) \to \varepsilon^2(k)$, and so (16.23).

16.24 *Thm.* $2 \leq x$ & $\exp(x, k, f)$ & $\exp(x, l, g)$ & $\exp(x, k \cdot l, h)$ & $j \leq l \to h(k \cdot j) \leq x \# 4 \# f(k) \# g(j)$.

Proof. Suppose $hyp\,(16.24)_j[0]$. Then $h(k \cdot 0) = 1$ and

$$x \# 4 \# f(k) \# g(0) = x \# 4 \# f(k) \# 1 = 1.$$

Thus $(16.24)_j[0]$. Suppose (16.24) & $hyp\,(16.24)_j[\mathrm{S}j]$. Let

$$a = x \# 4 \# f(k) \# g(j).$$

Then $h(k \cdot \mathrm{S}j) = h(k \cdot j) \cdot h(k) = h(k \cdot j) \cdot f(k) \leq a \cdot f(k)$. But

$$\begin{aligned} &x \# 4 \# f(k) \# g(\mathrm{S}j) \geq x \# 4 \# f(k) \# (2 \cdot g(j)) = \\ &a \cdot (x \# 4 \# f(k) \# 2) \geq a \cdot (2 \# 4 \# f(k) \# 2) = a \cdot (4 \# f(k)) = \\ &a \cdot |f(k)|_2^2 \geq a \cdot f(k). \end{aligned}$$

Thus $ind_j\,(16.24)$, and by BI we have (16.24).

16.25 *Thm.* $2 \leq x$ & $\exp(x, k, f)$ & $\exp(x, l, g) \to \exists h \exp(x, k \cdot l, h)$.

Proof. Let $a = x \# 4 \# f(k) \# g(l)$. By (16.24) and (16.11) we have (16.25): $h \leq \mathrm{K} \# (2 \cdot a) \# (2 \cdot a)$. Suppose $\exists k \exists f \neg(16.25)$. By BLNP there exist minimal such k and f. Clearly $k \neq 0$. As a consequence, we have $\exp(x, k-1, \mathrm{Rstr}(f, k-1))$, so there exists h_1 such that $\exp(x, (k-1) \cdot l, h_1)$. Since $\exp(x, l, g)$ and $k \cdot l = (k-1) \cdot l + l$, we conclude from (16.22) that $\exists h \exp(x, k \cdot l, h)$, a contradiction. Thus (16.25).

16.26 *Thm.* $\varepsilon(k) \leftrightarrow \varepsilon^3(k)$.

Proof. We have $\varepsilon^3(k) \rightarrow \varepsilon(k)$ by REL. Suppose $\varepsilon(k)$ & $\varepsilon(l)$. We claim that 1: $\exists h \exp(x, k \cdot l, h)$. Suppose $x < 2$. There exist f and g such that $\exp(2, k, f)$ & $\exp(2, l, g)$, so by (16.25) there exists h_1 such that $\exp(2, k \cdot l, h_1)$. By (16.21) we have (1), and thus $x < 2 \rightarrow (1)$. Suppose $2 \leq x$. There exist f and g such that $\exp(x, k, f)$ & $\exp(x, l, g)$, so by (16.23) we have $\varepsilon(k) \rightarrow \varepsilon^3(k)$, and so (16.26).

16.27 *Thm.* $\exp(2, k, f)$ & $\exp(x, k, g)$ & $i \leq k \rightarrow g(i) \leq (2 \cdot x) \# f(i)$.

Proof. Suppose $\exists i \neg(16.27)$. By BLNP there exists a minimal such i. Clearly $i \neq 0$. Consequently, $g(i-1) \leq (2 \cdot x) \# f(i-1)$, so

$$(2 \cdot x) \# f(i) = (2 \cdot x) \# (2 \cdot f(i-1)) =$$
$$|2 \cdot x|_2 \cdot ((2 \cdot x) \# f(i-1)) \geq x \cdot g(i-1) = g(i).$$

This is a contradiction, and thus (16.27).

16.28 *Thm.* $\exp(2, k, f)$ & $2 \leq x \rightarrow \exists g \exp(x, k, g)$.

Proof. Let $a = (2 \cdot x) \# f(k)$. By (16.27) and (16.11) we have (16.28): $g \leq \mathrm{K} \# (2 \cdot a) \# (2 \cdot a)$. Suppose $\exists k \exists f \neg(16.28)$. By BLNP there exist minimal such k and f. Clearly $k \neq 0$. We have $\exp(2, k-1, \mathrm{Rstr}(f, k-1))$, so there exists g_1 such that $\exp(x, k-1, g_1)$. Let $g = g_1 \cup \{\langle k, x \cdot g_1(k-1)\rangle\}$. Then *con* (16.28), a contradiction, and thus (16.28).

16.29 *Thm.* $\exp(2, k, f) \rightarrow \varepsilon(k)$.

Proof. By (16.28) and (16.21).

16.30 *Thm.* x is a power of two $\rightarrow$ x is a power of 2.

Proof. By (16.12), (16.29), and (16.26). □

This is a positive solution of the test problem of Chapter 14. It is more than just a test problem, because using it we can show that all of the nonlogical symbols of Chapter 14, except for ε and $\uparrow$, are equivalent to bounded symbols, and it will be essential in the sequel to have a bounded function symbol for the logarithm.

16.31 *Thm.* $\lambda(x)$.

Proof. By (16.30), $\exists k \; |x|_2 \cdot 2 = 2^k$. But $x \leq |x|_2 \cdot 2$, so (16.31).

16.32 *Def.* $\operatorname{Log} x = k \leftrightarrow \exists f(f \leq \mathrm{K} \# (2 \cdot |x|_2) \# (2 \cdot |x|_2)$ & $\exp(2, k, f)$ & $f(k) = |x|_2)$.

The existence condition holds by (16.30) and (16.11), and the uniqueness condition holds by (13.11).

16.33 *Thm.* $\operatorname{Log}_0 x = \operatorname{Log} x$.

Proof. By (14.11), (16.26), and (16.11).

16.34 *Thm.* $x \#_0 y = x \# y$.

Proof. Write α for

$$\exists f(f \leq \mathrm{K} \# (2 \cdot x \cdot y) \# (2 \cdot x \cdot y) \ \& \\ \exp(2, \operatorname{Log} x \cdot \operatorname{Log} y, f) \ \& \ f(\operatorname{Log} x \cdot \operatorname{Log} y) = x \# y).$$

Then (16.34) $\leftrightarrow \alpha$, by (16.33), (14.11), (14.3), (16.26), and (16.11). Suppose $\exists y \neg$(16.34). Then $\exists y \neg \alpha$, so by BLNP there exists y such that $\min_y \neg \alpha$, i.e., $\min_y x \#_0 y \neq x \# y$. Since $x \#_0 y = x \#_0 |y|_2$ and $x \# y = x \# |y|_2$, we have $y = |y|_2$ by the minimality assumption, so that y is a power of two. Clearly $y \neq 1$ and $y \neq 2$. Therefore $x \#_0 y = x \#_0 ((y/2) \cdot 2) = (x \#_0 (y/2)) \cdot (x \#_0 2) = (x \# (y/2)) \cdot (x \# 2) = x \# y$, again by the minimality assumption. This is a contradiction, and thus (16.34). □

Thanks to (16.31), (16.33), and (16.34), we will have no further use for the symbols λ, Log_0, and $\#_0$. The next result shows that "x is a power of b" is also equivalent to a bounded formula.

16.35 *Thm.* x is a power of $b \leftrightarrow (b = 0 \ \& \ (x = 0 \vee x = 1)) \vee (b = 1 \ \& \ x = 1)$ $\vee \ \exists k \exists f(k \leq x$ & $f \leq \mathrm{K} \# (2 \cdot x) \# (2 \cdot x)$ & $\exp(b, k, f)$ & $f(b) = x)$.

Proof. By (14.11), (16.26), and (16.11). □

We conclude this chapter by giving the defining axiom for superexponentiation. The uniqueness condition holds by (12.18).

16.36 *Def!* $x \Uparrow k = z \leftrightarrow \exists f(f$ is a function & $\forall i(i \in \operatorname{Dom} f \leftrightarrow i \leq k)$ & $f(0) = 1$ & $\forall i(i < k \leftrightarrow \varepsilon(f(i))$ & $f(i+1) = x \uparrow f(i))$ & $f(k) = z)$, otherwise $z = 0$.

Chapter 17

Bounded replacement

Let Q'_4 be the current theory.

Metatheorem 17.1 Let T *be an extension of* Q'_4, *and let* D *be such that* $\exists y\mathrm{D}$ *is a bounded formula of* T *and* $x_1, \ldots, x_\nu$ *are the variables distinct from* x *and* y *occurring free in* D. *Then*

$$\text{BR. } a \text{ is a set } \ \& \ \ \forall x(x \in a \rightarrow \exists y\mathrm{D}) \rightarrow \\ \exists f(f \text{ is a function } \ \& \ \ \mathrm{Dom}\, f = a \ \ \& \ \ \forall x(x \in a \rightarrow \mathrm{D}_y[f(x)]))$$

is a theorem of T, *and*

$$\text{BRD. } \{\langle x,y\rangle : x \in a \ \& \ \min_y \mathrm{D}\} = f \leftrightarrow f \text{ is a function } \ \& \\ \mathrm{Dom}\, f = a \ \ \& \ \ \forall x(x \in a \rightarrow (\min_y \mathrm{D})_y[f(x)]), \text{ otherwise } f = 1$$

is the defining axiom of a bounded function symbol. (The variables in the term $\{\langle x,y\rangle : x \in a \ \& \ \min_y \mathrm{D}\}$ *are* $a, x_1, \ldots, x_\nu$*.)*

Demonstration. Introduce the bounded function symbol f with defining axiom

$$\mathrm{f}xx_1 \ldots x_\nu = y \leftrightarrow \min_y \mathrm{D}, \text{ otherwise } y = 0.$$

We will suppress the variables $x_1, \ldots, x_\nu$ from now on, and we will write fx for f$x x_1 \ldots x_\nu$. To prove (BR), we argue in T as follows.

We claim that

$$1.\ \exists x(x \in a) \to \exists z(z \in a \ \&\ \forall w(w \in a \to \mathrm{f}w \le \mathrm{f}z)).$$

For suppose $\exists a \neg(1)$. By BLNP there exists a minimal such a. There exists x such that $x \in a$. Let $b = \{t \in a : t \ne x\}$. By (10.20), $b < a$; so by the minimality of a there exists z_1 such that $z_1 \in b \ \&\ \forall w(w \in b \to \mathrm{f}w \le \mathrm{f}z_1)$. Suppose $\mathrm{f}z_1 \le \mathrm{f}x$ and let $z = x$. Then we have a contradiction, and thus $\mathrm{f}x < \mathrm{f}z_1$. Let $z = z_1$. Again we have a contradiction, and thus (1).

Introduce the bounded function symbol $\mathrm{Maxm_f}$ with defining axiom

$$\mathrm{Maxm_f}\, a = z \leftrightarrow \min_z (z \in a \ \&\ \forall w(w \in a \to \mathrm{f}w \le \mathrm{f}z)),$$
$$\text{otherwise } z = 0,$$

and the bounded function symbol $\mathrm{Bd_f}$ by

$$\mathrm{Bd_f}\, a = |\mathrm{K} \cdot (\mathrm{SP\,Max}(a, \mathrm{f\,Maxm_f}\, a))^4|_2 \cdot 2.$$

Then we claim that

$$2.\ a \text{ is a set} \to \exists f(f \le (\mathrm{Bd_f}\, a) \# a \ \&\ f \text{ is a function } \&$$
$$\mathrm{Dom}\, f = a \ \&\ \forall x(x \in a \to f(x) = \mathrm{f}x)).$$

Suppose $\exists a \neg(2)$. By BLNP there exists a minimal such a. Suppose $a = 0$ and let $f = 0$. Then (2), a contradiction, and thus $a \ne 0$. Since a is a set, there exists x_0 such that $x_0 \in a$. Let $b = \{t \in a : t \ne x_0\}$. By (10.20), $2 \cdot b < a$. By the minimality of a there exists g such that $scope_{\exists f}(2)_{fa}[gb]$. Let $f = g \cup \{\langle x_0, \mathrm{f}x_0\rangle\}$. Then f is a function, $\mathrm{Dom}\, f = a$, and $\forall x(x \in a \to f(x) = \mathrm{f}x)$. By (16.5) we have

$$f \le \mathrm{K} \cdot ((\mathrm{Bd_f}\, b) \# b) \cdot (\mathrm{SP\,Max}(x_0, \mathrm{f}x_0))^4 \le (\mathrm{Bd_f}\, a) \cdot ((\mathrm{Bd_f}\, b) \# b).$$

By (1) we have $\mathrm{Bd_f}\, b \le \mathrm{Bd_f}\, a$, so by (16.8) we find

$$f \le (\mathrm{Bd_f}\, a) \cdot ((\mathrm{Bd_f}\, a) \# b) = (\mathrm{Bd_f}\, a) \# (2 \cdot b) \le (\mathrm{Bd_f}\, a) \# a.$$

Hence we have *con* (2), a contradiction, and thus (2).

Now suppose *hyp* (BR). There exists f such that $scope_{\exists f}$ (2). Suppose $x \in a$. Then $\exists y \mathrm{D}$, so by BLNP we have $\exists y \min_y \mathrm{D}$. Thus $x \in a \to \mathrm{D}_y[f(x)]$, and thus (BR).

The uniqueness condition for (BRD) holds by (12.2), and we have $\exists f\, rhs$ (BRD): $(\mathrm{Bd_f}\, a) \# a, x \le a$. □

Whenever a term of the form $\{\langle x, y\rangle : x \in a \ \&\ \min_y \mathrm{D}\}$ occurs, it is understood that the corresponding defining axiom of the form (BRD) has been adjoined to the theory. We will write

$$\{\langle x, y\rangle : x \in a \mathbin{\&} \mathrm{D}\} \quad \text{for} \quad \{\langle x, y\rangle : x \in a \mathbin{\&} \min_y \mathrm{D}\},$$

though we usually use this abbreviation only when $\vdash \mathrm{D} \leftrightarrow \min_y \mathrm{D}$. Also, we write

$$\{\langle x, \mathrm{b}\rangle : x \in a\} \quad \text{for} \quad \{\langle x, \mathrm{y}\rangle : x \in a \mathbin{\&} \mathrm{y} = \mathrm{b}\}$$

where y is the first variable in alphabetical order distinct from x and a and not occurring in b.

Chapter 18

An impassable barrier

Let us pause to examine from an impredicative point of view what we are doing. Take a strong theory T containing 0 and S, say an extension by definitions of Peano Arithmetic I or even of ZFC (Zermelo-Fraenkel set theory with the axiom of choice). Let $\hat{\mathrm{T}}$ be the theory obtained by adjoining a unary predicate symbol ϕ and the axiom

Fin. $\phi(0)$ & $(\phi(x) \to \phi(\mathrm{S}x))$.

This adjunction does not increase the power of the theory in any way; we can for example interpret $\phi(x)$ by $x = x$. If T is an axiomatization of arithmetic, say an extension by definitions of I, then by a *specific number* we mean a variable-free term b of T; if T is an extension by definitions of ZFC containing the constant ω denoting the set of all natural numbers, then by a *specific number* we mean a variable-free term b of T such that $\vdash_{\mathrm{T}} \mathrm{b} \in \omega$. We say that a specific number b is a *finite number* in case $\vdash_{\hat{\mathrm{T}}} \phi(\mathrm{b})$.

Now consider a specific number b, such as 10000 or $10 \uparrow 10 \uparrow 10$, and try to prove $\phi(\mathrm{b})$. The initial reaction of some mathematicians to such a problem is a failure to see the difficulty: they suggest proving $\phi(\mathrm{b})$ by induction. But even if the induction principle is an axiom scheme of T, we do not have induction available for $\phi(x)$, because the axioms of T say nothing about formulas containing ϕ. A second reaction is that the problem is trivial because there is an obvious proof in b steps. This observation manages to be both meaningless and incorrect; meaningless because of the ubiquitous pun confusing the formal and genetic concepts of number, and incorrect because there are specific numbers b for which one can show that there is no proof in $\hat{\mathrm{T}}$ of $\phi(\mathrm{b})$. That is, there is a specific number that is not a finite number.

To see this, we will let T be an extension by definitions of I. We assume that I is consistent; otherwise there is indeed a proof in $\hat{\mathrm{T}}$ of $\phi(\mathrm{b})$. Consider Gödel's construction of a closed formula $\forall n \mathrm{C}[n]$ that is unprovable even though $\mathrm{C}[0], \mathrm{C}[1], \mathrm{C}[2], \ldots$ are all provable. Let D be the unary formula $\mathrm{C}[n] \rightarrow \forall n\, C[n]$. Then $\exists n\, \mathrm{D}[n]$ is provable, since it is equivalent to the tautology $\forall n\, \mathrm{C}[n] \rightarrow \forall n\, \mathrm{C}[n]$, but $\mathrm{D}[0]$, $\mathrm{D}[1]$, $\mathrm{D}[2]$, $\ldots$ are all unprovable, since from a proof of one of them and the proof of the corresponding theorem among $\mathrm{C}[0]$, $\mathrm{C}[1]$, $\mathrm{C}[2]$, $\ldots$ we would immediately obtain a proof of $\forall n\, \mathrm{C}[n]$. Now consider the constant N with the defining axiom

$$1.\quad N = n \leftrightarrow \mathrm{D}[n] \;\&\; \forall m(m < n \rightarrow \neg\mathrm{D}[m]).$$

The existence condition holds by the least number principle and the uniqueness condition is obvious. We assume that (1) is an axiom of T. We claim that $\phi(N)$ is not a theorem of $\hat{\mathrm{T}}$. To see this, take a model of the consistent theory obtained by adjoining $\neg\mathrm{D}[0], \neg\mathrm{D}[1], \neg\mathrm{D}[2], \ldots$ to T and represent ϕ by membership in the smallest subset of the universe of the model containing the individual representing 0 and closed under the function representing S. Then $\phi(N)$ is not valid in the model, so $\phi(N)$ is not a theorem of $\hat{\mathrm{T}}$. I am grateful to Simon Kochen for this example.

Now consider a specific number b of the form $\mathrm{f}(0,\ldots,0)$, where f is a function symbol representing in T the primitive recursive function F. All mathematicians apparently believe that it is meaningful to speak of the number ρ such that if $\bar{\rho}$ is the term $\mathrm{S}\cdots\mathrm{S}0$ with ρ occurrences of S, then $\vdash_{\mathrm{T}} \bar{\rho} = \mathrm{b}$. To the Cantorian, ρ exists in a Platonic world; the intuitionist is convinced that $\bar{\rho}$ can in principle be constructed; the finitist accepts ρ because it is finite. But the argument that b is a finite number because there is a proof of $\phi(\mathrm{b})$ in ρ steps is circular—the number ρ is itself given as $F(0,\ldots,0)$, and how do we know that the putative proof will ever terminate? Consider the example that b is $\mathrm{SS}0 \Uparrow \mathrm{SSSSS}0$. What does it mean to speak of the term $\mathrm{S}\cdots\mathrm{S}0$ with $2 \Uparrow 5$, or $2 \uparrow 2 \uparrow 2 \uparrow 2 \uparrow 2$, or 2^{65536} occurrences of S? This involves the genetic concept of number. But if one produces occurrences of S at the rate of one every 10^{-24} seconds, which is about the time it takes light to traverse the diameter of a proton, and if the age of the universe is taken to be twenty billion years, then it will take more than 10^{19684} ages of the universe before $2 \Uparrow 5$ occurrences of S have been produced (and by the same token, what genetic meaning can 10^{19684} have?). The point is that to regard $2 \Uparrow 5$ as standing for a genetic number entails a philosophical commitment to some ideal notion of existence. To a nominalist, $2 \Uparrow 4$ stands for a number, 65536, to which one can count, but $2 \Uparrow 5$ is a pair of arabic numerals with a vertical double arrow between them, and there is not a scintilla of evidence that it stands for a genetic number. There is a story of a bank employee who was told to count a bundle of bill to verify that there was actually a thousand of them. The employee began to count them: $1, 2, 3, \ldots, 61, 62, 63$—and then stopped, being convinced that since it had checked perfectly all that way it must be correct.

There are more efficient ways of using the axiom (Fin) than step by step as a rule of inference, and we can actually prove $\phi(\mathrm{b})$ for specific numbers b much bigger than those we can actually count to. For example, let

$$\mathrm{a}_0 = 2,\ \mathrm{a}_1 = \mathrm{a}_0 \cdot \mathrm{a}_0,\ \mathrm{a}_2 = \mathrm{a}_1 \cdot \mathrm{a}_1,\ \ldots,\ \mathrm{a}_{16} = \mathrm{a}_{15} \cdot \mathrm{a}_{15},$$

so that $\vdash \mathrm{a}_{16} = 2 \Uparrow 5$. Form $\phi^3(x)$. Then $\phi^3(x)$ respects multiplication and is stronger than $\phi(x)$, so one quickly obtains proofs of

$$\phi^3(\mathrm{a}_0),\ \phi^3(\mathrm{a}_1),\ \phi^3(\mathrm{a}_2),\ \ldots,\ \phi^3(\mathrm{a}_{16}),\ \phi^3(2 \Uparrow 5),$$

and so of $\phi(2 \Uparrow 5)$. Encouraged by this success, one might expect that for every construction of a function symbol f representing a primitive recursive function, one could constuct an inductive predicate symbol ϕ' stronger than ϕ to be used in proving $\phi'(\mathrm{f}(0, \ldots, 0))$ and so $\phi(\mathrm{f}(0, \ldots, 0))$, but this is not so. The reader may enjoy seeing how to construct proofs of the formulas $\phi(2 \Uparrow 6), \phi(2 \Uparrow 7), \ldots, \phi(2 \Uparrow 2 \Uparrow 4)$ in extensions by definition of $\hat{\mathrm{T}}$. But there are limits to such methods; no one will ever prove $\phi(2 \Uparrow 2 \Uparrow 5)$. Proofs ordinarily involve the use of quantifiers, and Hilbert and Ackermann showed how to eliminate quantifiers from a proof. When all quantifiers have been eliminated, we are reduced to using (Fin) step by step as a rule of inference, and so a proof of $\phi(\mathrm{b})$ with quantifiers eliminated must have at least ρ steps, where $\vdash_{\mathrm{T}} \bar{\rho} = \mathrm{b}$. But one can estimate the increase in the length of a proof when quantifiers are eliminated, and so obtain a lower bound on the length of the original proof. As one can see from this description, the argument involves accepting rather large formal numbers as standing for genetic numbers. The arguments given in this chapter to establish Assertions 18.1 and 18.2 are finitary, but no claim is made that they can be carried out predicatively.

By the *rank* of a formula we mean the number of occurrences of $\exists$ in it. In the following Assertions we will use *proof* (in a theory T) to mean a sequence of formulas of T each of which is an axiom of T, is a tautological consequence of preceding formulas, or can be inferred by $\exists$-introduction from a preceding formula (see the discussion of induction on theorems in [Sh,§3.1]), and a proof *of* A is a proof whose last formula is A. By the *rank* of a proof we mean the maximal rank of the formulas in it. Let Z be the theory whose nonlogical symbols are 0 and S, and whose nonlogical axioms are (3.1) and (3.2); i.e., $\mathrm{S}x \neq 0$ and $\mathrm{S}x = \mathrm{S}y \rightarrow x = y$. Then $\vdash_Z \bar{p}_1 \neq \bar{p}_2$ whenever p_1 is unequal to p_2. We use various arithmetical symbols metamathematically, including $\uparrow$ for exponentiation and $\Uparrow$ for superexponentiation.

Assertion 18.1 Let T *be a consistent open extension of Z, and let* b *be a variable-free term of* T *such that* $\vdash_{\mathrm{T}} \mathrm{b} \neq \bar{p}$ *for all p with* $p < 2 \Uparrow \mu$. *Then any proof in* $\hat{\mathrm{T}}$ *of* $\phi(\mathrm{b})$ *of rank τ has at least* $2 \Uparrow (\mu - 2 \cdot \tau - 3)$ *formulas.*

Argument. Refer to the proof of the Consistency Theorem in [Sh,§4.3]. Assume that we are given a proof in $\hat{\mathrm{T}}$ of $\phi(\mathrm{b})$ of rank τ with η formulas. By Lemma 1 of [Sh,§4.3], $\phi(\mathrm{b})$ is a tautological consequence of formulas in $\Delta(\hat{\mathrm{T}})$. Let us examine the proof of Lemma 1 to see how many are required. Let F be the smallest function such that for any proof with λ formulas of a formula A in a theory U, any closed instance of A is a tautological consequence of $F(\lambda)$ formulas in $\Delta(\mathrm{U})$. Then we have from the proof of Lemma 1 that

$$F(1) = 1 \quad \text{and} \quad F(\lambda) \leq \max\left(\sum_{\beta=1}^{\lambda-1} F(\beta), 1 + \max_{\beta<\lambda} F(\beta)\right)$$

(according as A is an axiom, is a tautological consequence of preceding formulas, or is inferred by $\exists$-introduction from a preceding formula). Hence $F(\lambda) \leq 2^{\lambda} - 1$, and so $\phi(\mathrm{b})$ is a tautological consequence of $2^{\eta-1}$ formulas in $\Delta(\hat{\mathrm{T}})$. Let U be the theory $\hat{\mathrm{T}}[\neg\phi(\mathrm{b})]$; then U is an open inconsistent theory, and there is a special sequence (that is, a sequence of formulas such that the disjunction of their negations is a tautology) of 2^{η} formulas in $\Delta_\tau(\mathrm{U})$. We let $K(\nu)$ be the minimal number of formulas in a special sequence all of whose formulas are in $\Delta_\nu(\mathrm{U})$. Then $K(\tau) \leq 2^\eta$. Since $2 \Uparrow (\mu - 2 \cdot \tau - 3) = \mathrm{Log}(2 \Uparrow (\mu - 2 \cdot \tau - 2))$ and $\mathrm{Log}\, K(\tau) \leq \eta$, it suffices to show that $2 \Uparrow (\mu - 2 \cdot \tau - 2) \leq K(\tau)$.

We claim that $2 \Uparrow \mu < K(0)$. From a special sequence containing $K(0)$ formulas in $\Delta_0(\mathrm{U})$ we obtain, upon deleting the instances of identity axioms and equality axioms, a quasi-tautology $\neg \mathbf{A}_1 \vee \cdots \vee \neg \mathbf{A}_\lambda$, where each $\mathbf{A}_\alpha$ for α from 1 to λ is an instance of a nonlogical axiom of U, and $\lambda \leq K(0)$. Since T is consistent, $\hat{\mathrm{T}}$ is consistent, because we can interpret $\phi(x)$ by $x = x$. Therefore the axiom $\neg\phi(\mathrm{b})$ of U must be one of the $\mathbf{A}_\alpha$. Let the other formulas among the $\mathbf{A}_\alpha$ that contain ϕ be $(\mathrm{Fin})[\mathrm{b}_1], \ldots, (\mathrm{Fin})[\mathrm{b}_\gamma]$, so that $\gamma < K(0)$, and let T_0 be

$$\mathrm{T}[(\mathrm{Fin})[\mathrm{b}_1], \ldots, (\mathrm{Fin})[\mathrm{b}_\gamma]].$$

Then we have

1. $\vdash_{\mathrm{T}_0} \phi(\mathrm{b})$.

Let B_ρ be the formula $\mathrm{b}_1 = \bar{\rho} \vee \cdots \vee \mathrm{b}_\gamma = \bar{\rho}$. Since T is a consistent extension of Z, at most γ of the formulas B_ρ are theorems of T. Suppose $\gamma < 2 \Uparrow \mu$. Then there is a ρ with $\rho < 2 \Uparrow \mu$ such that we can consistently adjoin $\neg\mathrm{B}_\rho$ to T. But then there is an interpretation I of T_0 in $\mathrm{T}[\neg\mathrm{B}_\rho]$ in which

$$\phi_{\mathrm{I}}(x) \text{ is } x = 0 \vee x = \mathrm{S}0 \vee \cdots \vee x = \bar{\rho};$$

in fact, for $\beta = 1, \ldots, \gamma$ we have

$$\vdash_{\mathrm{T}[\neg\mathrm{B}_\rho]} \mathrm{b}_\beta = 0 \vee \mathrm{b}_\beta = \mathrm{S}0 \vee \cdots \vee \mathrm{b}_\beta = \bar{\rho} \rightarrow \\ \mathrm{Sb}_\beta = 0 \vee \mathrm{Sb}_\beta = \mathrm{S}0 \vee \cdots \vee \mathrm{Sb}_\beta = \bar{\rho}$$

since $\vdash_{\mathrm{T}[\neg\mathrm{B}_\rho]} \mathrm{b}_\beta \neq \bar{\rho}$. Using (1) we obtain a proof in $\mathrm{T}[\neg\mathrm{B}_\rho]$ of $\phi(\mathrm{b})$, i.e. of $\mathrm{b} = 0 \vee \mathrm{b} = \mathrm{S}0 \vee \cdots \vee \mathrm{b} = \bar{\rho}$, which is impossible. Thus $2 \Uparrow \mu \leq \gamma < K(0)$, which establishes the claim.

Consider a special sequence of κ formulas. In the proof of Lemma 2 of [Sh,§4.3], the special sequence is divided into three portions: $\mathbf{A}_i$ for $i = 1, \ldots, r$; $\mathrm{B_x}[\mathrm{a}_j] \rightarrow \exists\mathrm{xB}$ for $j = 1, \ldots, p$; and $\exists\mathrm{xB} \rightarrow \mathrm{B_x}[\mathrm{r}]$, so that $r + p + 1 = \kappa$. From this a new special sequence is obtained (with the formulas belonging to the special constant r eliminated), and the number of formulas in it is $r \cdot (p+1)$. Since $r + (p+1) = \kappa$, this is at most $\mathrm{Qt}(4, \kappa^2)$. The next time we eliminate formulas belonging to a special constant we get at most $\mathrm{Qt}(4, \kappa^4)$ (in fact, at most $\mathrm{Qt}(4^3, \kappa^4)$) formulas, and when we have eliminated all formulas belonging to special constants of maximal rank we get at most $\mathrm{Qt}(4, \kappa \uparrow 2 \uparrow \kappa)$ formulas. That is, $4 \cdot K(\nu - 1) \le K(\nu) \uparrow 2 \uparrow K(\nu)$. Let

$$L(\nu) = K(\nu) + \mathrm{Log}(\mathrm{Log}\, K(\nu) + 1) + 1.$$

Then $4 \cdot K(\nu - 1) \le 2 \uparrow 2 \uparrow L(\nu - 1)$ and $L(\nu - 1) \le 2 \uparrow 2 \uparrow L(\nu)$. We have already seen that $2 \Uparrow \mu < K(0)$. Therefore $2 \Uparrow (\mu - 1) < L(0)$, so

$$2 \Uparrow (\mu - 3) < L(1), \ldots, 2 \Uparrow (\mu - 2 \cdot \tau - 1) < L(\tau),$$

and hence $2 \Uparrow (\mu - 2 \cdot \tau - 2) < K(\tau)$, as was to be shown. □

Earlier I said rather dogmatically that no one will ever prove the formula $\phi(2 \Uparrow 2 \Uparrow 5)$. Assertion 18.1 gives some evidence for this. Suppose we try to prove $\phi(2 \Uparrow 2 \Uparrow 5)$ starting from a consistent extension T of Z to which we adjoin (Fin). The theory T may not be open, but by Skolem's Theorem (see [Sh,§4.5]) it has a conservative open extension T^0, and each axiom of T can be proved rather quickly from the corresponding axiom of T^0. However, in Assertion 18.1 we used a restricted notion of proof. There are various devices, discussed in Chapters 3 and 4 of [Sh], that shorten a proof: use of extensions by definition, the Deduction Theorem, special constants, and so forth. One has the impression on reading about these devices that they shorten proofs by at most exponential factors, which are quite negligible in comparison to the bound of Assertion 18.1. This impression is worth studying in detail; we will do so later on. For the present, let me summarize by saying that I can count to $2 \Uparrow 4$ but not $2 \Uparrow 5$, and I can prove $\phi(2 \Uparrow 2 \Uparrow 4)$ but not $\phi(2 \Uparrow 2 \Uparrow 5)$.

Now we turn to the consequences of Assertion 18.1 for the predicative theory of arithmetic that we are developing. The reader may have wondered why we did not extend the relativization scheme to incorporate $\uparrow$ instead of #. If one looks at the proof of REL in Chapter 5, the natural next step would seem to be to let $\hat{\mathrm{C}}^4$ be an abbreviation for $\forall y(\mathrm{C}^3[y] \to \mathrm{C}^3[y \uparrow x])$, and then try to establish the further claims

16. $\hat{\mathrm{C}}^4[x] \rightarrow \hat{\mathrm{C}}^3[x]$,
17. $\hat{\mathrm{C}}^4[0]$,
18. $\hat{\mathrm{C}}^4[x] \rightarrow \hat{\mathrm{C}}^4[\mathrm{S}x]$,
19. $\hat{\mathrm{C}}^4[x]$ & $u \leq x \rightarrow \hat{\mathrm{C}}^4[u]$,
20. $\hat{\mathrm{C}}^4[x_1]$ & $\hat{\mathrm{C}}^4[x_2] \rightarrow \hat{\mathrm{C}}^4[x_1 + x_2]$,
21. $\hat{\mathrm{C}}^4[x_1]$ & $\hat{\mathrm{C}}^4[x_2] \rightarrow \hat{\mathrm{C}}^4[x_1 \cdot x_2]$,
22. $\hat{\mathrm{C}}^4[x_1]$ & $\hat{\mathrm{C}}^4[x_2] \rightarrow \hat{\mathrm{C}}^4[x_1 \uparrow x_2]$.

The trouble in trying to establish (22) is that $\uparrow$ (unlike $+$, $\cdot$, and $\#$) is not associative. We cannot in this way extend the relativization scheme to incorporate $\uparrow$. This raises the question as to whether some other scheme might work, but the following Assertion is that there is no relativization scheme incorporating $\uparrow$ for the generic inductive formula ϕ. See also [PD].

Assertion 18.2 Let T *be a consistent extension of* Q_2'. *1: No inductive formula of* $\hat{\mathrm{T}}$ *is stronger than* $\phi(x)$ *and respects* $\uparrow$. *2: No inductive formula of* $\hat{T}$ *is stronger than* $\phi(2 \Uparrow x)$.

Argument. Suppose C_1 is an inductive formula of $\hat{\mathrm{T}}$ that is stronger than $\phi(x)$ and respects $\uparrow$, and let C_2 be $\mathrm{C}_1[2 \Uparrow x]$. It is easy to see that $2 \Uparrow 0 = 1$, $2 \Uparrow x \neq 0 \rightarrow 2 \Uparrow \mathrm{S}x = 2 \uparrow (2 \Uparrow x)$, and $2 \Uparrow x = 0 \rightarrow 2 \Uparrow \mathrm{S}x = 0$ are theorems of Q_4'. Therefore C_2 is inductive and stronger than $\phi(2 \Uparrow x)$. Thus to establish (1) it suffices to establish (2).

Suppose C_2 is an inductive formula of $\hat{\mathrm{T}}$ that is stronger than $\phi(2 \Uparrow x)$. Let T^0 be an open conservative extension of T; this exists by Skolem's Theorem (see [Sh,§4.5]). The formula C_2^3 respects addition, so that

$$\vdash_{\widehat{\mathrm{T}^0}} \mathrm{C}_2^3[x] \rightarrow \mathrm{C}_2^3[x + x].$$

Hence the number of formulas in a proof in $\widehat{\mathrm{T}^0}$ of $\phi(2 \Uparrow (2 \uparrow \bar{\rho}))$ grows only linearly with ρ while the rank remains constant, which contradicts Assertion 18.1. Thus Assertion 18.2 holds. □

This result may serve to deepen one's mistrust of the induction principle. There is a barrier separating the predicative from the impredicative, a barrier as absolute as that between the genetic and the formal, and more sharply delineated than the debatable demarcation between the finite and the infinite.

As syntactical methods for establishing certain inductions have failed, we may be tempted to turn to semantics and appeal to the set ω of all natural numbers. Of course it is a theorem of axiomatic set theory that formalized Peano Arithmetic is consistent, but this is not what people have in mind who argue that Peano Arithmetic is consistent because its axioms are true statements about ω. What is at issue here is not the familiar construct of formal mathematics, but a belief in the existence of ω prior to all mathematical constructions. What is the origin of this belief? The famous saying by Kronecker that God created the numbers, all else is the work of Man, presumably was not meant to be taken seriously. Nowhere in the book of Genesis do we find the passage: And God said, let there be numbers, and there were numbers; odd and even created he them, and he said unto them, be fruitful and multiply; and he commanded them to keep the laws of induction. No, the belief in ω stems from the speculations of Greek philosophy on the existence of ideal entities or the speculations of German philosophy on a priori categories of thought. An appeal to ω to justify the induction principle is no more secure than are these philosophical systems, and yet it is hard to relinquish. We are creatures (Kronecker had it backwards), not too much older than an infant in a crib, and we still feel the urge to count on something when we count. The infant counts on its fingers, the mathematician counts on ω—but the infant at least knows its fingers to exist. The mathematician's attitude towards ω has in practice been one of faith, and faith in a hypothetical entity of our own devising, to which are ascribed attributes of necessary existence and infinite magnitude, is idolatry.

We now have an extensive development of predicative arithmetic and some idea of its limitations. These limitations were found using the Consistency Theorem. The proof of the Consistency Theorem is finitary; in fact, it was devised with the laudable aim of defending formal mathematics from the attacks of the intuitionists. But can the Consistency Theorem itself be established predicatively? More generally, which results of mathematical logic can be established predicatively? Using our stringent test of predicativity as interpretability in Q, we can study this question by arithmetizing syntax within our predicative theory of arithmetic and seeking predicatively to prove arithmetizations of the fundamental results of finitary mathematical logic. Let us do this, following [Sh] in shameless detail. It is a long journey. Before setting out we may remark that it is doomed to failure. The Consistency Theorem can be used to prove the consistency of Q, so if we could establish, within our theory, both it and Gödel's theorem on the impossibility of self-consistency proofs, then we would have a contradiction. This indicates that at least one of these two pillars of finitary mathematical logic, the Hilbert-Ackermann Consistency Theorem and Gödel's Second Theorem, makes an appeal to impredicative concepts.

In addition to philosophical objections to impredicative methods, one may have mathematical misgivings about their use. Perhaps the place to look for possible trouble is not in the upper branches of set theory, but rather at the very roots in arithmetic where impredicativity first appears.

Chapter 19

Sequences

In this chapter we will predicatively arithmetize the basic syntactical notions of juxtaposition, substitution, and occurrence.

19.1 *Thm.* $\exists z \forall w\, f(w) \leq f(z)$.

Proof. We have (19.1): $z \leq f, w \leq f$. Suppose $\exists f \neg$(19.1). By BLNP there exists a minimal such f. Suppose $\neg(f$ is a function$) \vee f = 0$, and let $z = 0$. Then (19.1), a contradiction, and thus f is a function & $f \neq 0$. By (12.2) there exists t such that $t \in f$, and there exist x and y such that $\langle x, y \rangle = t$. Let $g = \{s \in f : s \neq t\}$. By the minimality of f there exists z_1 such that $\forall w\, g(w) \leq g(z_1)$. Suppose $f(z_1) \leq f(x)$ and let $z = x$. Then (19.1), a contradiction, and thus $f(x) < f(z_1)$. Let $z = z_1$. Then (19.1), a contradiction, and thus (19.1).

19.2 *Def.* $\operatorname{Maxm} f = z \leftrightarrow \min_z \forall w\, f(w) \leq f(z)$.

The existence condition holds by (19.1) and BLNP, the uniqueness condition is obvious, and we have $\exists z\, rhs$ (19.2): $z \leq f, w \leq f$.

19.3 *Def.* $\operatorname{Sup} f = f(\operatorname{Maxm} f)$.

19.4 *Def.* u is a sequence $\leftrightarrow$ u is a function & $\exists n \forall i (i$ is in the domain of $u \leftrightarrow 1 \leq i \leq n)$.

19.5 *Def.* $\operatorname{Ln} u = n \leftrightarrow u$ is a sequence & $\forall i(i$ is in the domain of $u \leftrightarrow 1 \leq i \leq n)$, otherwise $n = 0$.

19.6 *Def.* $\operatorname{Chop} u = \{z \in u : z \neq \langle \operatorname{Ln} u, u(\operatorname{Ln} u)\rangle\}$.

19.7 *Thm.* u is a sequence & $u \neq 0 \to \text{Chop}\, u$ is a sequence & $\text{Ln Chop}\, u + 1 = \text{Ln}\, u$ & $2 \cdot \text{Chop}\, u < u$.

Proof. Suppose *hyp*(19.7). Then $\text{Chop}\, u$ is a function and $\forall i(i$ is in the domain of $\text{Chop}\, u \leftrightarrow 1 \le i \le \text{Ln}\, u - 1)$, so $\text{Chop}\, u$ is a sequence and $\text{Ln Chop}\, u + 1 = \text{Ln}\, u$. We have $2 \cdot \text{Chop}\, u < u$ by (10.20). Thus (19.7).

19.8 *Thm.* $\text{Ln}\, u \le \text{Log}\, u$.

Proof. Suppose $\exists u \neg(19.8)$. By BLNP there exists a minimal such u. Suppose $\neg(u$ is a sequence$) \vee u = 0$. Then $\text{Ln}\, u = 0$ and so (19.8), a contradiction. Thus u is a sequence & $u \neq 0$. By the minimality of u and (19.7), $\text{Ln Chop}\, u \le \text{Log Chop}\, u$. Since $2 \cdot \text{Chop}\, u < u$ we have $\text{Ln}\, u = \text{Ln Chop}\, u + 1 \le \text{Log Chop}\, u + 1 < \text{Log}\, u$. Hence (19.8), a contradiction, and thus (19.8).

19.9 *Thm.* $\exists a(a$ is a set & $\forall i(i \in a \leftrightarrow i < \text{Log}\, x)$ &
$a \le (730 \cdot (\text{Log}\, x)^2) \# x)$.

Proof. Suppose $x = 0$ and let $a = 0$. Then (19.9), and thus $(19.9)_x[0]$. Suppose (19.9). Then there exists a such that $scope_{\exists a}(19.9)$. We have $\text{Log}\,(x+1) = \text{Log}\, x \to (19.9)_x[\text{S}x]$, so suppose $\text{Log}\,(x+1) \neq \text{Log}\, x$. Then $\text{Log}\,(x+1) = \text{Log}\, x + 1$ and $|x+1|_2 = |2 \cdot x|_2$. Let $a_1 = a \cup \{\text{Log}\, x\}$. Then a_1 is a set and $\forall i(i \in a_1 \leftrightarrow i < \text{Log}\,(x+1))$. Let $c = 730 \cdot (\text{Log}\,(x+1))^2$. By (16.2), (16.1), and (16.8),

$$
\begin{aligned}
&a_1 \le 5 \cdot \text{SP}a \cdot 146 \cdot (\text{Log}\, x)^2 \le \\
&730 \cdot (\text{Log}\, x)^2 \cdot ((730 \cdot (\text{Log}\, x)^2) \# x) \le \\
&|c|_2 \cdot (c \# x) = c \# (2 \cdot x) = c \# |2 \cdot x|_2 = c \# |x+1|_2 = c \# (x+1).
\end{aligned}
$$

Thus $(19.9)_x[\text{S}x]$, and thus $ind_x\,(19.9)$. Hence (19.9) by BI.

19.10 *Def.* $\text{Setlog}\, x = a \leftrightarrow a$ is a set & $\forall i(i \in a \leftrightarrow i \le \text{Log}\, x)$.

The uniqueness condition holds by (12.2), and by (19.9) we have the existence condition and $\exists a\, rhs\,(19.10)$: $a \le (730 \cdot (\text{Log}\,(x+1))^2) \# (x+1)$, $i \le \text{Max}(a, \text{Log}\, x)$.

19.11 *Def.* $\text{Explogfn}(x, y) = f \leftrightarrow \exp(x, \text{Log}\, y, f)$.

The existence condition holds by (14.11) and the uniqueness condition holds by (13.2). Suppose $\exp(x, \text{Log}\, y, f)$. Let $z = \text{Max}(x, 2)$. There exists g such that $\exp(z, \text{Log}\, y, g)$. We have $f \le g$ by (16.20) and we have $g \le \text{K} \# (2 \cdot g(\text{Log}\, y)) \# (2 \cdot g(\text{Log}\, y))$ by (16.11). But

$$g(\mathrm{Log}\,y) = z \uparrow \mathrm{Log}\,y \leq 2 \uparrow ((\mathrm{Log}\,z + 1) \cdot \mathrm{Log}\,y) \leq (2 \cdot z) \# y = \mathrm{Max}(2 \cdot x, 4) \# y.$$

Thus we obtain that $\exists f\, rhs\,(19.11)\colon f \leq \mathrm{K} \# (2 \cdot (\mathrm{Max}(2 \cdot x, 4) \# y)) \# (2 \cdot (\mathrm{Max}(2 \cdot x, 4) \# y))$.

19.12 *Def.* $\mathrm{Explog}(x, y) = \mathrm{Explogfn}(x, y)(\mathrm{Log}\,y)$.

19.13 *Thm.* $\mathrm{Explog}(x, y) = x^{\mathrm{Log}\,y}$.

Proof. By (16.26). □

The point is that although the term $x \uparrow \mathrm{Log}\,y$ contains the unbounded function symbol $\uparrow$, it can be written as a bounded function symbol applied to x and y.

19.14 *Def.* $\mathrm{Expln}(x, y) = \mathrm{Explogfn}(\mathrm{Ln}\,y)$.

19.15 *Thm.* $\mathrm{Expln}(x, y) = x^{\mathrm{Ln}\,y}$.

Proof. By (19.8).

19.16 *Thm.* u is a sequence $\rightarrow u \leq (\mathrm{K} \cdot (\mathrm{Ln}\,u \cdot \mathrm{SP}\,\mathrm{Sup}\,u)^4)^{\mathrm{Ln}\,u}$.

Proof. We have $con\,(19.16) \leftrightarrow u \leq \mathrm{Expln}(\mathrm{K} \cdot (\mathrm{Ln}\,u \cdot \mathrm{SP}\,\mathrm{Sup}\,u)^4, u)$. Suppose $\exists u \neg(19.16)$. By BLNP there exists u such that $\min_u \neg(19.16)$. (I hope that no one will object that $(\ )^4$ involves an unbounded function symbol!) Clearly $u \neq 0$. Let $n = \mathrm{Ln}\,u$ and let $v = \mathrm{Chop}\,u$. Then $v \leq (\mathrm{K} \cdot ((n-1) \cdot \mathrm{SP}\,\mathrm{Sup}\,u)^4)^{n-1}$. We have $u = v \cup \{\langle n, u(n) \rangle\}$, so by (16.5) we have $con\,(19.16)$, a contradiction. Thus (19.16).

19.17 *Thm.* u is a sequence & $\mathrm{Ln}\,u \leq \mathrm{Log}\,y$ & $\mathrm{Sup}\,u \leq z \rightarrow u \leq (2 \cdot \mathrm{K} \cdot (\mathrm{Log}\,y \cdot \mathrm{SP}z)^4) \# y$.

Proof. Suppose $hyp\,(19.17)$ and let $w = \mathrm{K} \cdot (\mathrm{Log}\,y \cdot \mathrm{SP}z)^4$. By (19.16) we have $u \leq w \uparrow \mathrm{Log}\,y \leq 2 \uparrow ((\mathrm{Log}\,w + 1) \cdot \mathrm{Log}\,y) \leq (2 \cdot w) \# y$, and thus (19.17). □

Let u_μ be either a unary function symbol or the empty expression, for all μ from 1 to ν. We use

$$\mathrm{A}\colon \mathrm{u}_1\mathrm{x}_1 \leq \mathrm{a}_1, \ldots, \mathrm{u}_\nu\mathrm{x}_\nu \leq \mathrm{a}_\nu$$

as an abbreviation for $\mathrm{A} \leftrightarrow \mathrm{A}'$, where A' is the formula obtained by replacing each part of A of the form $\exists \mathrm{x}_\mu \mathrm{B}$ by $\exists \mathrm{x}_\mu(\mathrm{u}_\mu \mathrm{x}_\mu \leq \mathrm{a}_\mu \;\&\; \mathrm{B})$, for all μ from

1 to ν. If all the u_μ are the empty expression, this agrees with our previous notation (see Chap. 7).

If x occurs in A only in parts of the form ∃x(x is a sequence & B)—or more generally, if A is equivalent to the formula obtained by replacing each part of A of the form ∃xB by ∃x(x is a sequence & B)—and if we have

$$\mathrm{A}: \mathrm{Ln}\,\mathrm{x} \leq \mathrm{Log}\,\mathrm{a}, \mathrm{Sup}\,\mathrm{x} \leq \mathrm{b}$$

or

$$\mathrm{A}: \mathrm{Ln}\,\mathrm{x} \leq \mathrm{Ln}\,\mathrm{a}, \mathrm{Sup}\,\mathrm{x} \leq \mathrm{b},$$

then it follows from (19.17) and (19.8) that

$$\mathrm{A}: \mathrm{x} \leq (2 \cdot \mathrm{K} \cdot (\mathrm{Log}\,\mathrm{a} \cdot \mathrm{SPb})^4) \# \mathrm{a}.$$

In showing a formula to be bounded, we will usually write one of the former, leaving the latter to be inferred.

Now we introduce juxtaposition.

19.18 *Def.* $u * v = \{\langle i, y\rangle : i \in \mathrm{Setlog}\,(u \cdot v)$ & $1 \leq i \leq \mathrm{Ln}\,u + \mathrm{Ln}\,v$ & $(1 \leq i \leq \mathrm{Ln}\,u \to y = u(i))$ & $(\mathrm{Ln}\,u + 1 \leq i \leq \mathrm{Ln}\,u + \mathrm{Ln}\,v \to y = v(i - \mathrm{Ln}\,u))\}$.

19.19 *Thm.* $u * v$ is a sequence & $\mathrm{Ln}\,(u * v) = \mathrm{Ln}\,u + \mathrm{Ln}\,v$.

Proof. Clearly $u * v$ is a function. We have $\forall i(i$ is in the domain of $u * v \leftrightarrow 1 \leq i \leq \mathrm{Ln}\,u + \mathrm{Ln}\,v)$ since $\mathrm{Ln}\,u + \mathrm{Ln}\,v \leq \mathrm{Log}\,(u \cdot v)$ by (19.8).

19.20 *Thm.* $(u * v) * w = u * (v * w)$.

Proof. Let $l = (u * v) * w$ and let $r = u * (v * w)$. We claim that 1: $l(i) = r(i)$. Suppose 2: $1 \leq i \leq \mathrm{Ln}\,u$. Then $l(i) = (u * v)(i) = u(i)$ and $r(i) = u(i)$. Thus (2) → (1). Suppose 3: $\mathrm{Ln}\,u + 1 \leq i \leq \mathrm{Ln}\,u + \mathrm{Ln}\,v$. Then $l(i) = (u * v)(i) = v(i - \mathrm{Ln}\,u)$ and $r(i) = (v * w)(i - \mathrm{Ln}\,u) = v(i - \mathrm{Ln}\,u)$, and thus (3) → (1). Suppose 4: $\mathrm{Ln}\,u + \mathrm{Ln}\,v + 1 \leq i \leq \mathrm{Ln}\,u + \mathrm{Ln}\,v + \mathrm{Ln}\,w$. Then we obtain $l(i) = w(i - (\mathrm{Ln}\,u + \mathrm{Ln}\,v)) = w(i - \mathrm{Ln}\,u - \mathrm{Ln}\,v)$ and $r(i) = (v * w)(i - \mathrm{Ln}\,u) = w(i - \mathrm{Ln}\,u - \mathrm{Ln}\,v)$, and thus (4) → (1). Suppose 5: $i = 0 \vee \mathrm{Ln}\,u + \mathrm{Ln}\,v + \mathrm{Ln}\,w < i$. Then $l(i) = 0$ and $r(i) = 0$, and thus (5) → (1). Hence (1), and by (12.2) we have (19.20).

19.21 *Def.* $\mathrm{sum}(u, v) \leftrightarrow u$ and v are sequences & $\mathrm{Ln}\,u = \mathrm{Ln}\,v$ & $v(1) = u(1)$ & $\forall i(1 \leq i < \mathrm{Ln}\,u \to v(i + 1) = v(i) + u(i + 1))$.

19.22 *Thm.* u is a sequence $\rightarrow \exists v(\text{sum}(u,v) \ \& \ \text{Sup}\, v \leq \text{Ln}\, u \cdot \text{Sup}\, u)$.

Proof. We have (19.22): $\text{Ln}\, v \leq \text{Ln}\, u, \text{Sup}\, v \leq \text{Ln}\, u \cdot \text{Sup}\, u$. Suppose $\exists u \neg(19.22)$. By BLNP there exists a minimal such u. Suppose $u = 0$ and let $v = 0$. Then (19.22), a contradiction, and thus $u \neq 0$. Let $n = \text{Ln}\, u$ and let $u_1 = \text{Chop}\, u$. There exists v_1 such that $\text{sum}(u_1, v_1) \ \& \ \text{Sup}\, v_1 \leq (n-1) \cdot \text{Sup}\, u_1$. Let $v = v_1 \cup \{\langle n, v_1(n-1) + u(n)\rangle\}$. Then $\text{sum}(u,v)$. We have

$$\text{Sup}\, u = \text{Max}(\text{Sup}\, v_1, v_1(n-1) + u(n)) \leq \\ (n-1) \cdot \text{Sup}\, u + \text{Sup}\, u = n \cdot \text{Sup}\, u,$$

so that (19.22), a contradiction, and thus (19.22).

19.23 *Def.* $\sum u = v \leftrightarrow \text{sum}(u,v)$, otherwise $v = 0$.

The uniqueness condition holds by (12.17), and we have

$$\exists v\, rhs\, (19.23)\text{: } \text{Ln}\, v \leq \text{Ln}\, u, \text{Sup}\, v \leq \text{Ln}\, u \cdot \text{Sup}\, v.$$

Notice that $\sum u$ is the sequence of partial sums, and the total sum is $(\sum u)(\text{Ln}\, u)$.

19.24 *Def.* $\text{prod}(u,v) \leftrightarrow u$ and v are sequences $\ \& \ \text{Ln}\, u = \text{Ln}\, v \ \& \ v(1) = u(1) \ \& \ \forall i(1 \leq i < \text{Ln}\, u \rightarrow v(i+1) = v(i) \cdot u(i+1))$.

19.25 *Thm.* u is a sequence $\rightarrow \exists v(\text{prod}(u,v) \ \& \ \text{Sup}\, v \leq \text{Expln}(\text{Sup}\, u, u))$.

Proof. Suppose $\exists u \neg(19.25)$. By BLNP there exists a minimal such u. Suppose $u = 0$ and let $v = 0$. Then (19.25), a contradiction, and thus $u \neq 0$. Let $n = \text{Ln}\, u$ and let $u_1 = \text{Chop}\, u$. Then there exists v_1 such that $\text{prod}(u_1, v_1) \ \& \ \text{Sup}\, v_1 \leq (\text{Sup}\, u_1)^{n-1}$. Let $v = v_1 \cup \{\langle n, v_1(n-1) \cdot u(n)\rangle\}$. Then $\text{prod}(u,v)$. We have

$$\text{Sup}\, v = \text{Max}(\text{Sup}\, v_1, v_1(n-1) \cdot u(n)) \leq (\text{Sup}\, u)^{n-1} \cdot \text{Sup}\, u = (\text{Sup}\, u)^n,$$

so that (19.25), a contradiction, and thus (19.25).

19.26 *Def.* $\prod u = v \leftrightarrow \text{prod}(u,v)$, otherwise $v = 0$.

19.27 *Thm.* $\varepsilon(n) \leftrightarrow \exists a(a \text{ is a set } \ \& \ \forall i(i \in a \leftrightarrow i \leq n))$.

Proof. Suppose $\varepsilon(n)$. Then there exists f such that $\exp(2, n, f)$. We have $n = \text{Log}\, f(n)$. Let $a = \text{Setlog}\, f(n)$. Then $rhs\, (19.27)$, and thus

$\varepsilon(n) \to rhs\,(19.27)$. Conversely, suppose a is a set & $\forall i(i \in a \leftrightarrow i \leq n)$. Let $u = \{\langle i,x\rangle : i \in a\}$ and let $f = \prod u$. Then $\exp(x,n,f)$, and thus (19.27). □

Let *** be obtained from — by forming the plural of a noun in —, let a be a term, and let x be the first variable in alphabetical order not occurring in — or a. Then we write "a is a sequence of ***" for

$$\text{a is a sequence} \;\;\&\;\; \forall \mathrm{x}(1 \leq \mathrm{x} \leq \mathrm{Ln\,a} \to \mathrm{a(x)} \text{ is a } \text{—}).$$

Also, we write "a is a set of ***" for

$$\text{a is a set} \;\;\&\;\; \forall \mathrm{x}(\mathrm{x} \in \mathrm{a} \to \mathrm{x} \text{ is a } \text{—}).$$

Let f be a bounded unary function symbol and let a be a bounded term. Then we write

$$\mathrm{f} \circ \mathrm{a} \;\; \text{for} \;\; \{\langle \mathrm{x}, \mathrm{f}(\mathrm{a}(\mathrm{x}))\rangle : \mathrm{x} \in \mathrm{Dom\,a}\}.$$

Now we shall introduce the juxtaposition of a sequence of sequences. The following two function symbols locate the sequence, and the index in that sequence, corresponding to an index in the juxtaposition.

19.28 *Def.* $\mathrm{Loc}_1(i,s) = j \leftrightarrow \min_j i \leq (\sum \mathrm{Ln} \circ s)(j)$, otherwise $j = 0$.

We have $\exists j\, rhs\,(19.28)$: $j \leq \mathrm{Ln}\, s$.

19.29 *Def.* $\mathrm{Loc}_2(i,s) = i - (\sum \mathrm{Ln} \circ s)(\mathrm{Loc}_1(i,s) - 1)$.

19.30 *Def.* $s^* = \{\langle i,y\rangle : i \in \mathrm{Setlog}\,(s(\mathrm{Maxm}\,(\mathrm{Ln} \circ s)) \# s) \;\;\&\;\; 1 \leq i \leq (\sum \mathrm{Ln} \circ s)(\mathrm{Ln}\, s) \;\;\&\;\; y = s(\mathrm{Loc}_1(i,s))(\mathrm{Loc}_2(i,s))\}$.

19.31 *Thm.* s is a sequence of sequences $\to s^*$ is a sequence & $\mathrm{Ln}\, s^* = (\sum \mathrm{Ln} \circ s)(\mathrm{Ln}\, s)$.

Proof. Suppose $hyp\,(19.31)$. Clearly s^* is a function. We have

$$(\textstyle\sum \mathrm{Ln}\, s)(\mathrm{Ln}\, s) \leq \mathrm{Sup}\,(\mathrm{Ln} \circ s) \cdot \mathrm{Ln}\, s = \mathrm{Ln}\,(s(\mathrm{Maxm}\,(\mathrm{Ln} \circ s))) \cdot \mathrm{Ln}\, s \leq \mathrm{Log}\,(s(\mathrm{Maxm}\,(\mathrm{Ln} \circ s))) \cdot \mathrm{Log}\, s = \mathrm{Ln}\,(s(\mathrm{Maxm}\,(\mathrm{Ln} \circ s)) \# s)$$

by (14.20), and so $\forall i(i$ is in the domain of $s^* \leftrightarrow 1 \le i \le (\sum \mathrm{Ln} \circ s)(\mathrm{Ln}\, s))$. Thus (19.31).

19.32 *Def.* $\langle x \rangle = \{\langle 1, x \rangle\}$.

Now we introduce substitution and simultaneous substitution.

19.33 *Def.* $\mathrm{Sub}(u, x, v) = \{\langle i, y \rangle : i \in \mathrm{Dom}\, u \;\; \& \;\; (\langle u(i) \rangle \ne x \to y = \langle u(i) \rangle) \;\; \& \;\; (\langle u(i) \rangle = x \to y = v)\}^*$.

19.34 *Def.* $\mathrm{Ssub}(u, w, s) = \{\langle i, y \rangle : i \in \mathrm{Dom}\, u \;\; \& \;\; \neg \exists j(1 \le j \le \mathrm{Ln}\, w \;\; \& \;\; \langle u(i) \rangle = w(j) \to y = \langle u(i) \rangle) \;\; \& \;\; \forall j(\min_j (1 \le j \le \mathrm{Ln}\, w \;\; \& \;\; \langle u(i) \rangle = w(j) \to y = s(j)))\}^*$.

19.35 *Def.* $u[i, j] = \{\langle k, u(i+k-1) \rangle : k \in \{k \in \mathrm{Dom}\, u : 1 \le k \le j-i+1\}\}$.

19.36 *Def.* v occurs in $u \leftrightarrow \exists i \exists j(1 \le i \le j \le \mathrm{Ln}\, u \;\; \& \;\; v = u[i, j])$.

19.37 *Thm.* v occurs in $u \to v \le u$.

Proof. Suppose v occurs in u. There exist i and j such that $1 \le i \le j \le \mathrm{Ln}\, u \;\; \& \;\; v = u[i, j]$. Let

$$f = \{\langle z, w \rangle : z \in v \;\&\; w = \langle i + \mathrm{Proj}_1\, z - 1, \mathrm{Proj}_2\, z \rangle\}.$$

Then we have $\forall z(z \in v \to z \le f(z))$ by (16.19). By (12.19) we have $v \le u$. Thus (19.37). □

Notice that now that we have bounded replacement, we can dispense with FS (Metatheorem 12.1) and use (12.19) instead.

The proof of (19.37) relies on the fact (16.19) that $\langle x, y \rangle$ is increasing in both x and y. This property fails for the usual set-theoretic definition of ordered pair—it takes a sudden dip when $x = y$. The result (19.37) will often be used tacitly in arguments by BLNP.

19.38 *Thm.* u is a sequence $\;\&\;\; 1 \le i \le \mathrm{Ln}\, u \to u[1, i-1] * u[i+1, \mathrm{Ln}\, u] < u$.

Proof. Suppose *hyp* (19.38). Let $v = u[1, i-1] * u[i+1, \mathrm{Ln}\, u]$ and let

$$f = \{\langle z, w \rangle : z \in v \;\&\; (\mathrm{Proj}_1\, z \le i - 1 \to w = z) \;\&\; (i \le \mathrm{Proj}_1\, z \to w = \langle \mathrm{Proj}_1\, z + 1, \mathrm{Proj}_2\, z \rangle)\}.$$

Then we have $\forall z(z \in v \to z \le f(z))$ by (16.19). By (12.19) we have $v \le u$. Clearly $v \ne u$, so $v < u$. Thus (19.38).

19.39 *Thm.* w occurs in v & v occurs in $u \to w$ occurs in u.

Proof. Suppose *hyp* (19.39). There exist i, j, k, and l such that $v = u[i,j]$ & $w = v[k,l]$. Let $i_1 = i + k - 1$ and let $j_1 = i + l - 1$. Then $w = u[i_1, j_1]$. Thus (19.39).

19.40 *Def.* Reverse $u = \{\langle i, u(\text{Ln}\, u + 1 - i)\rangle : i \in \text{Dom}\, u\}$.

Chapter 20

Cardinality

A set has to be small in three ways: the formula describing it must be bounded, there must be a bound on its elements, and there must be a logarithmic bound on its cardinality. We elaborate on this statement in this chapter and the next.

20.1 *Def.* f is injective $\leftrightarrow \forall x \forall y(x \in \operatorname{Dom} f \ \& \ \ y \in \operatorname{Dom} f \ \ \& \ \ x \neq y \rightarrow f(x) \neq f(y))$.

We use the abbreviations "a is an injective sequence" for "a is injective & a is a sequence", and "a is an injective sequence of ***" for "a is injective & a is a sequence of ***".

20.2 *Def.* u is an injection into $a \leftrightarrow u$ is an injective sequence & $\operatorname{Ran} u \subseteq a$.

20.3 *Thm.* u is an injection into $a \rightarrow \operatorname{Ln} u \leq \operatorname{Log} a$.

Proof. Suppose $\exists u \exists a \neg(20.3)$. By BLNP there exist minimal such u and a. Clearly $u \neq 0$. Let $a_1 = \{x \in a : x \neq u(\operatorname{Ln} u)\}$. Then $2 \cdot a_1 < a$ by (10.20), and $\operatorname{Chop} u$ is an injection into a_1, so $\operatorname{Ln} \operatorname{Chop} u \leq \operatorname{Log} a_1$ and hence $\operatorname{Ln} u \leq \operatorname{Log} a$, a contradiction. Thus (20.3).

20.4 *Def.* $\operatorname{Card} a = n \leftrightarrow \min_n \forall u(u$ is an injection into $a \rightarrow \operatorname{Ln} u \leq n)$.

We have $\exists n \, rhs\,(20.4)$: $n \leq \operatorname{Log} a$, $\operatorname{Ln} u \leq \operatorname{Log} a$, $\operatorname{Sup} u \leq a$. The existence condition holds by (20.3) and BLNP, and the uniqueness condition is obvious.

20.5 *Thm.* $\operatorname{Card} a \leq \operatorname{Log} a$.

Proof. By (20.3).

20.6 *Thm.* $\exists u(u$ is an injection into a & $\operatorname{Ln} u = \operatorname{Card} a)$.

Proof. From (20.4).

20.7 *Thm.* $x \in a$ & $a_1 = \{y \in a : y \neq x\} \to \operatorname{Card} a = \operatorname{Card} a_1 + 1$.

Proof. Suppose *hyp* (20.7), suppose 1: u is an injection into a, and suppose 2: $1 \leq j \leq \operatorname{Ln} u$ & $u(j) = x$. Let

$$v = \{\langle i, y\rangle : i \in \operatorname{Dom} u \;\&\; 1 \leq i \leq \operatorname{Ln} u - 1 \;\&\; (i < j \to y = u(i)) \;\&\; (j \leq i \to y = u(i+1))\}.$$

Then v is an injection into a_1 and hence $\operatorname{Ln} v \leq \operatorname{Card} a_1$, so that 3: $\operatorname{Ln} u \leq \operatorname{Card} a_1 + 1$. Thus (2) $\to$ (3). Suppose 4: $\forall j(1 \leq j \leq \operatorname{Ln} u \to u(j) \neq x)$. Then u is an injection into a_1, so that $\operatorname{Ln} u \leq \operatorname{Card} a_1$. Thus (4) $\to$ (3), and thus (1) $\to$ (3). Therefore $\operatorname{Card} a \leq \operatorname{Card} a_1 + 1$. By (20.6) there exists u_1 such that u_1 is an injection into a_1 & $\operatorname{Ln} u_1 = \operatorname{Card} a_1$. Let $u = u_1 \cup \{\langle \operatorname{Ln} u_1 + 1, x\rangle\}$. Then u is an injection into a and $\operatorname{Ln} u = \operatorname{Card} a_1 + 1$, so $\operatorname{Card} a_1 + 1 \leq \operatorname{Card} a$. Thus (20.7).

20.8 *Def.* $\operatorname{Bd} a = \operatorname{Max} x\,(x \leq a \;\&\; x \in a)$.

20.9 *Def.* $\operatorname{Expcard}(x, a) = \operatorname{Explogfn}(x, a)(\operatorname{Card} a)$.

20.10 *Thm.* $\operatorname{Expcard}(x, a) = x^{\operatorname{Card} a}$.

Proof. By (20.5).

20.11 *Thm.* a is a set $\to a \leq \operatorname{Expcard}(730 \cdot (\operatorname{SP} \operatorname{Bd} a)^2, a)$.

Proof. Suppose $\exists a \neg(20.11)$. By BLNP there exists a minimal such a. Clearly $a \neq 0$, so there exists x such that $x \in a$. Let $a_1 = \{y \in a : y \neq x\}$. Then $a_1 \leq (730 \cdot (\operatorname{SP} \operatorname{Bd} a_1)^2) \uparrow \operatorname{Card} a$. But $a \leq 730 \cdot \operatorname{SP} a_1 \cdot (\operatorname{SP} x)^2$ by (16.2) and (16.1), so that

$$a \leq (730 \cdot (\operatorname{SP} \operatorname{Bd} a)^2) \cdot (730 \cdot (\operatorname{SP} \operatorname{Bd} a)^2)^{\operatorname{Card} a_1} = (730 \cdot (\operatorname{SP} \operatorname{Bd} a)^2)^{\operatorname{Card} a}$$

by (20.7). This is a contradiction, and thus (20.11).

20.12 *Thm.* f is a function $\to$ Card f = Card Dom f.

Proof. Suppose $\exists f \neg(20.12)$. By BLNP there exists a minimal such f. Clearly $f \neq 0$, so there exists z such that $z \in f$. Let $a = \operatorname{Dom} f$, let $f_1 = \{t \in f : t \neq z\}$, and let $a_1 = \{s \in a : s \neq \operatorname{Proj}_1 z\}$. Then f_1 is a function, $\operatorname{Dom} f_1 = a_1$, and $f_1 < f$, so $\operatorname{Card} f_1 = \operatorname{Card} a_1$. But by (20.7) we have $\operatorname{Card} f = \operatorname{Card} f_1 + 1$ and $\operatorname{Card} a = \operatorname{Card} a_1 + 1$, so $\operatorname{Card} f = \operatorname{Card} a$, a contradiction. Thus (20.12).

20.13 *Thm.* $a \cap b = 0 \to \operatorname{Card}(a \cup b) = \operatorname{Card} a + \operatorname{Card} b$.

Proof. Suppose $\exists b \neg(20.13)$. By BLNP there exists a minimal such b. Suppose $\neg\exists x\, x \in b$. Then $a \cup b = a$ and $\operatorname{Card} b = 0$, so *con* (20.13), a contradiction. Thus there exists x such that $x \in b$. Let $b_1 = \{t \in b : t \neq x\}$. Then $a \cap b_1 = 0$, so $\operatorname{Card}(a \cup b_1) = \operatorname{Card} a + \operatorname{Card} b_1$. By (20.7) we have $\operatorname{Card}(a \cup b) = \operatorname{Card}(a \cup b_1) + 1$ and $\operatorname{Card} b = \operatorname{Card} b_1 + 1$, so *con* (20.13), a contradiction. Thus (20.13).

20.14 *Thm.*
$$\begin{aligned}\exists c(c \text{ is a set } \ \&\ & \\ & \forall z(z \in c \leftrightarrow \exists x \exists y(x \in a \ \&\ y \in b \ \&\ z = \langle x, y\rangle)) \ \& \\ & \operatorname{Card} c = \operatorname{Card} a \cdot \operatorname{Card} b \ \& \\ & c \leq \operatorname{Explog}(\mathrm{K} \cdot (\operatorname{Max}(\operatorname{Bd} a, \operatorname{Bd} b))^4, a \# b)).\end{aligned}$$

Proof. Suppose $\exists b \neg(20.14)$. By BLNP there exists a minimal such b. Suppose $\neg\exists y_0\, y_0 \in b$ and let $c = 0$. Then (20.14), a contradiction, and thus there exists y_0 such that $y_0 \in b$. Let $b_1 = \{t \in b : t \neq y_0\}$. Then there exists c_1 such that $scope_{\exists c}\,(20.14)_{cb}[c_1 b_1]$. Let $f = \{\langle x, y_0\rangle : x \in a\}$ and let $c = c_1 \cup f$. Now $c_1 \cap f = 0$, so we have $\operatorname{Card} c = \operatorname{Card} a \cdot \operatorname{Card} b_1 + \operatorname{Card} f = \operatorname{Card} a \cdot \operatorname{Card} b_1 + \operatorname{Card} a = \operatorname{Card} a \cdot \operatorname{Card} b$ by (20.13), (20.12), and (20.7). Clearly c is a set and $scope_{\forall z}\,(20.14)$. Let $k = (\operatorname{Max}(\operatorname{Bd} a, \operatorname{Bd} b))^4$. By (16.3) we have $\operatorname{Bd} c \leq 5 \cdot (\operatorname{Max}(\operatorname{Bd} a, \operatorname{Bd} b))^2$, so that

$$\begin{aligned}&c \leq (730 \cdot 25 \cdot k) \uparrow \operatorname{Card} c = (\mathrm{K} \cdot k) \uparrow (\operatorname{Card} a \cdot \operatorname{Card} b) \leq \\ &(\mathrm{K} \cdot k) \uparrow (\operatorname{Log} a \cdot \operatorname{Log} b) = (\mathrm{K} \cdot k) \uparrow \operatorname{Log}(a \# b) = \operatorname{Explog}(\mathrm{K} \cdot k, a \# b)\end{aligned}$$

by (20.11), (20.5), and (14.20). Therefore we have (20.14), a contradiction, and thus (20.14).

20.15 *Def.* $a \times b = c \leftrightarrow c$ is a set & $\forall z(z \in c \leftrightarrow \exists x \exists y(x \in a \ \& \ y \in b \ \& \ z = \langle x, y \rangle))$.

The uniqueness condition holds by (12.2) and the existence condition and boundedness hold by (20.14).

20.16 *Def.* $\mathrm{Occ}\, u = \mathrm{Ran}\, \{\langle k, v \rangle : k \in \mathrm{Dom}\, u \times \mathrm{Dom}\, v \ \& \ \exists i \exists j (1 \le i \le j \le \mathrm{Ln}\, u \ \& \ k = \langle i, j \rangle \ \& \ u[i,j] = v)\}$.

20.17 *Thm.* v occurs in $u \leftrightarrow v \in \mathrm{Occ}\, u$.

Proof. From (20.16).

20.18 *Thm.* f is a function & $a = \mathrm{Dom}\, f \rightarrow$
$f \le \mathrm{Expcard}(\mathrm{K} \cdot (\mathrm{SP}\,\mathrm{Max}(\mathrm{Bd}\, a, \mathrm{Sup}\, f))^4, a)$.

Proof. Suppose *hyp* (20.18). It follows from (20.12) and (20.11) that $f \le (730 \cdot (\mathrm{SP}\,\mathrm{Bd}\, f)^2) \uparrow \mathrm{Card}\, a$, so by (16.1) and (16.3) we have *con* (20.18). Thus (20.18). □

If x occurs in A only in parts of the form $\exists$x(x is a function & B)—or more generally, if A is equivalent to the formula obtained by replacing each part of A of the form $\exists$xB by $\exists$x(x is a function & B)—and if we have

$$\mathrm{A} \colon \mathrm{Dom}\,\mathrm{x} \le \mathrm{a}, \mathrm{Sup}\,\mathrm{x} \le \mathrm{b},$$

then it follows from (20.18) and (20.5) that we have

$$\mathrm{A} \colon \mathrm{x} \le \mathrm{Explog}(\mathrm{K} \cdot (\mathrm{SP}\,\mathrm{Max}(\mathrm{a}, \mathrm{b}))^4, \mathrm{a}).$$

Similarly, if x occurs in A only in parts of the form $\exists$x(x is a set & B)—or more generally, if A is equivalent to the formula obtained by replacing each part of A of the form $\exists$xB by $\exists$x(x is a set & B)—and if we have

$$\mathrm{A} \colon \mathrm{Bd}\,\mathrm{x} \le \mathrm{a}, \mathrm{Card}\,\mathrm{x} \le \mathrm{Log}\,\mathrm{b},$$

then it follows from (20.11) and (20.5) that we have

$$\mathrm{A} \colon \mathrm{x} \le \mathrm{Explog}(730 \cdot (\mathrm{SPa})^2, \mathrm{b}).$$

In showing a formula to be bounded, we will usually write one of the former, leaving the bound on x to be inferred.

The following result shows that if we have a function r on a set (think of r as a ranking function), then we can enumerate the elements of the set in such a way that those of lower rank come first.

20.19 *Thm.* a is a set $\to \exists u(u$ is an injection into a & $\operatorname{Ran} u = a$ & $\forall i \forall j(1 \le i \le \operatorname{Ln} u$ & $1 \le j \le \operatorname{Ln} u$ & $r(u(i)) < r(u(j)) \to i < j))$.

Proof. We have (20.19): $\operatorname{Ln} u \le \operatorname{Log} a$, $\operatorname{Sup} u \le a$ by (20.3). Suppose $\exists a \neg$(20.19). By BLNP there exists a minimal such a. Clearly $a \ne 0$. Let $r_0 = \{z \in r : \operatorname{Proj}_1 z \in a\}$ and let $x_1 = \operatorname{Maxm} r_0$. Then $x_1 \in a$. Let $a_1 = \{t \in a : t \ne x_1\}$. By the minimality assumption there exists u_1 such that $scope_{\exists u}$ $(20.19)_{ua}[u_1 a_1]$. Let $u = u_1 \cup \{\langle \operatorname{Ln} u_1 + 1, x \rangle\}$. Then (20.19), a contradiction, and thus (20.19).

20.20 *Def.* $\operatorname{Enumer}(a, r) = u \leftrightarrow \min_u scope_{\exists u}$ (20.19).

We have $\exists u$ *rhs* (20.20): $\operatorname{Ln} u \le \operatorname{Log} a$, $\operatorname{Sup} u \le \operatorname{Bd} a$. The existence condition holds by (20.19) and BLNP, and the uniqueness condition is obvious.

20.21 *Def.* $\operatorname{Enum} a = \operatorname{Enumer}(a, 0)$.

Chapter 21

Existence of sets

Metatheorem 21.1 Let Q_4'' *be the current theory, let* U *be an extension of* Q_4''*, let* A *be a bounded formula of* U*, and let* a *and* b *be bounded terms of* U *not containing* x *or* y*. Then the following is a theorem of* U:

$$\text{SET. } \forall x(A \to x \leq a) \ \&\ \forall y(\forall x(x \in y \to A) \to \operatorname{Card} y \leq \operatorname{Log} b) \to \exists y(y \text{ is a set } \&\ \forall x(x \in y \leftrightarrow A)).$$

Demonstration. We prove (SET) in U as follows. Suppose *hyp*(SET). Write α for

$$y \leq \operatorname{Explog}(730 \cdot (\mathrm{SPa})^2, b) \ \&\ y \text{ is a set } \&\ \forall x(x \in y \to A),$$

and let $y = \operatorname{Max} y\, \alpha$. Clearly $\alpha_y[0]$, so by MAX we have α. Suppose $A \ \&\ x \notin y$ and let $y_1 = y \cup \{x\}$. (Here y_1 is distinct from x and y and does not occur in A, a, or b.) By (20.11) we have $\alpha_y[y_1]$, a contradiction, and thus $\forall x(x \in y \leftrightarrow A)$. Thus (SET). □

We have developed a certain amount of set theory, but it is only a small portion of Cantorian set theory; see Figure 21.1. Different mathematicians who study the foundations of mathematics incorporate different portions of set theory into their metamathematical belief system. Platonists discuss whether measurable cardinals exist, but they believe in R_1. In some sense intuitionists believe in ω while finitists believe only in the elements of R_ω, though it is difficult to make a strict comparison.

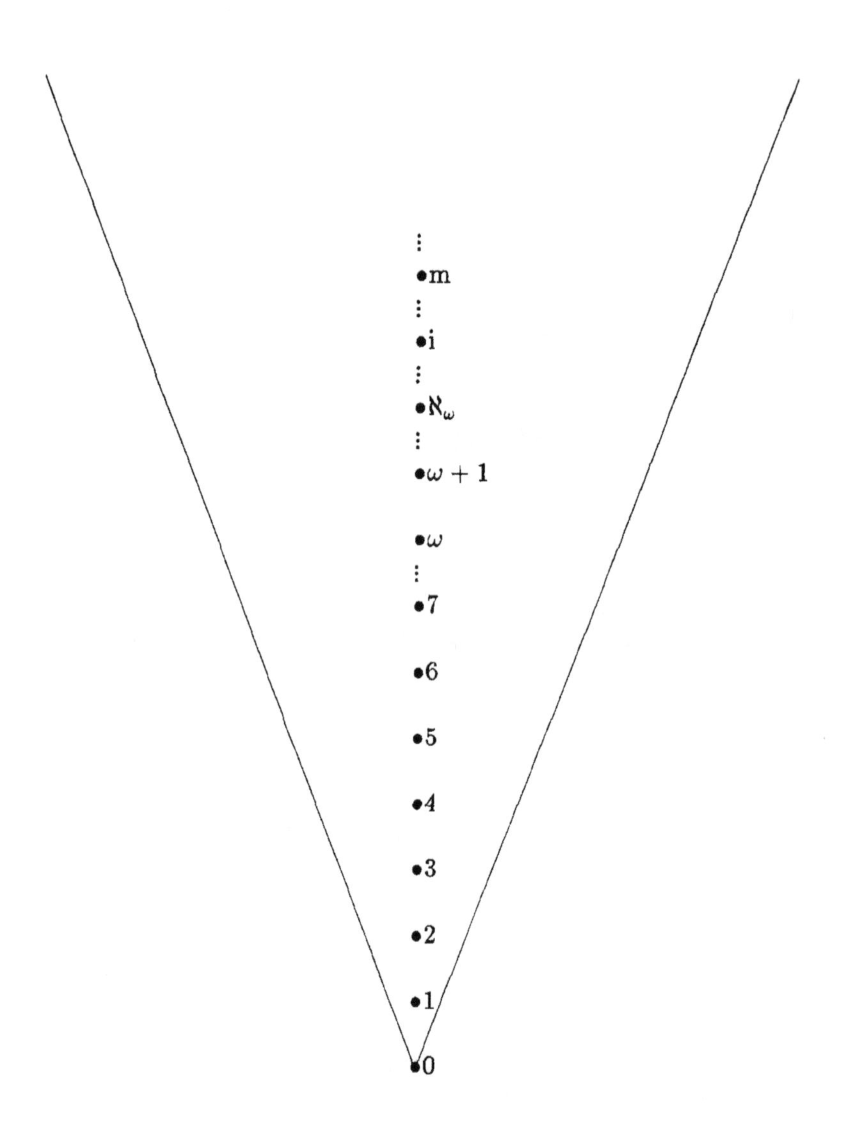

The ordinals are depicted on the vertical axis, and the portion of the figure lying strictly below the horizontal line through the ordinal α *depicts the set* R_α *of all sets of rank* $< \alpha$*. Here* i *is the first inaccessible cardinal and* m *is the first measurable cardinal. The figure is not drawn to scale.*

Figure 21.1: Cantor's Paradise

Set theory is abstract, but abstract beliefs affect concrete actions. Someone who believes in R_1 would consider it a waste of time to try to produce a contradiction from the axioms of ZFC; similarly for $R_{\omega \cdot 2}$ and Zermelo's original axiomatic set theory with separation but without replacement; and similarly for $R_{\omega+1}$ and Peano Arithmetic.

To a nominalist it is clear where to draw the line separating the real from the speculative: R_5, which is a system of 65536 objects, exists—but R_6, with its $2 \Uparrow 5$ members, is only a formal construct.

Chapter 22

Semibounded replacement

This chapter is a digression, and I do not intend to use it in the sequel except in occasional remarks. The semibounded replacement principle differs from the bounded replacement principle of Chapter 17 in that $\exists$yD is no longer required to be bounded, though D itself is.

Metatheorem 22.1 Let D *be a bounded formula of* Q_4'' *such that* $x_1, \ldots, x_\nu$ *are the variables distinct from* x *and* y *occurring free in* D*. Consider the formula*

$$\text{SBR. } a \text{ is a set } \ \& \ \forall x(x \in a \rightarrow \exists y\mathrm{D}) \rightarrow \\ \exists f(f \text{ is a function } \ \& \ \operatorname{Dom} f = a \ \& \ \forall x(x \in a \rightarrow \mathrm{D}_y[f(x)])).$$

Then $Q_4''[(\mathrm{SBR})]$ *is interpretable in* Q_4''.

Demonstration. Let D_1 be $\min_y \mathrm{D}$. We want to construct the function f with domain a such that for $x \in a$ we have $f(x) = y$ if and only if D_1. How do we know such a function exists? Write $\varsigma[n]$ for

$$\forall a \forall x_1 \cdots \forall x_\nu (a \text{ is a set } \ \& \ \operatorname{Card} a \leq n \ \& \ \forall x(x \in a \rightarrow \exists y\mathrm{D}) \rightarrow \\ \exists f(f \text{ is a function } \ \& \ \operatorname{Dom} f = a \ \& \ \forall x(x \in a \rightarrow \mathrm{D}_1))).$$

We claim that $ind_n\, \varsigma[n]$. Clearly $\varsigma[0]$. Suppose

$$\varsigma[n] \ \& \ a \text{ is a set } \ \& \ \operatorname{Card} a = n+1 \ \& \ \forall x(x \in a \rightarrow \exists y\mathrm{D}).$$

There exists z such that $z \in a$. Let $a_0 = \{t \in a : t \neq z\}$. Then there exists f_0 such that f_0 is a function & $\operatorname{Dom} f_0 = a_0$ & $\forall x(x \in a_0 \to \mathrm{D}_1)$. Since $z \in a$ we have $\exists w\, \mathrm{D}_{xy}[zw]$, so by BLNP there exists w such that $\mathrm{D}_{1xy}[zw]$. Let $f = f_0 \cup \{\langle z, w\rangle\}$. Thus $\varsigma[n] \to \varsigma[n+1]$, and the claim is established.

Since ς is inductive, we can form ς^4 and apply SREL of Chapter 15. Then we have $\varsigma^4[a] \to \varsigma^4[\operatorname{Card} a]$, and so $\varsigma^4[a] \to \varsigma[\operatorname{Card} a]$, since Card is a bounded function symbol. Therefore we have

$$\begin{aligned}&\varsigma^4[a] \ \& \ a \text{ is a set } \& \ \forall x(x \in a \to \exists y \mathrm{D}) \to \\ &\exists f(f \text{ is a function } \& \ \operatorname{Dom} f = a \ \& \ \forall x(x \in a \to \mathrm{D}_1))\end{aligned}$$

and a fortiori

$$\begin{aligned}1.\ &\varsigma^4[a] \ \& \ a \text{ is a set } \& \ \forall x(x \in a \to \exists y(\varsigma^4[y] \ \& \ \mathrm{D})) \to \\ &\exists f(f \text{ is a function } \& \ \operatorname{Dom} f = a \ \& \ \forall x(x \in a \to \mathrm{D}_1)).\end{aligned}$$

Suppose *hyp* (1). Then there exists f such that $\mathit{scope}_{\exists f}\, \mathit{con}$ (1). By (20.18) we have

$$f \leq \operatorname{Expcard}(\mathrm{K} \cdot (\mathrm{SP} \operatorname{Max}(\operatorname{Bd} a, \operatorname{Sup} f))^4, a).$$

But $\operatorname{Sup} f = f(\operatorname{Maxm} f)$. Let $z = \operatorname{Maxm} f$. Then $z \in a$, so $\exists w(\varsigma^4[w] \ \& \ \mathrm{D}_{xy}[zw])$, so there exists w such that $\varsigma^4[w]$ & $\mathrm{D}_{1xy}[zw]$. That is, $\varsigma^4[f(z)]$ and so $\varsigma^4[\operatorname{Sup} f]$. Consequently we have $\varsigma^4[f]$, and thus

$$\begin{aligned}2.\ &\varsigma^4[a] \ \& \ a \text{ is a set } \& \ \forall x(x \in a \to \exists y(\varsigma^4[y] \ \& \ \mathrm{D})) \to \\ &\exists f(\varsigma^4[f] \ \& \ f \text{ is a function } \& \ \operatorname{Dom} f = a \ \& \ \forall x(x \in a \to \mathrm{D})).\end{aligned}$$

But (2) is equivalent to the relativization of (SBR) by ς^4.

Therefore we have an interpretation I of $Q_4''[(\mathrm{SBR})]$ in Q_4'' constructed as follows. The universe $\mathrm{U_I}$ of I is given by $\mathrm{U_I} n \leftrightarrow \varsigma^4[n]$. For each bounded symbol u of the theory, $\mathrm{u_I}$ is u. For each unbounded symbol u of the theory, $\mathrm{u_I}$ is the symbol whose defining axiom is given by the relativization by ς^4 of the defining axiom of u. □

We can always introduce a function symbol by means of the defining axiom

$$\text{SBRD!}\ \{\langle x,y\rangle : x \in a\ \&\ \min_y \mathrm{D}\} = f \leftrightarrow f \text{ is a function } \&$$
$$\operatorname{Dom} f = a\ \&\ \forall x(x \in a \to (\min_y \mathrm{D})_y[f(x)]), \text{ otherwise } f = 1,$$

even if D is unbounded, since the uniqueness condition holds. If D is bounded and we have SBR, such defining axioms are useful.

The theory obtained by adjoining all axioms of the form (SBR), with D bounded, is locally interpretable in Q, but we will continue to work only in extensions by definition of Q_4.

Chapter 23

Formulas

Now we are ready to begin to investigate which results of finitary mathematical logic can be established predicatively. We will follow the presentation in [Sh] very closely. See [Sh,§2.4] in connection with this chapter.

It will be convenient to enlarge our stock of variables. We also let A, B, C, and D- possibly with 0, 1, 2, ... as a subscript—be variables, and if x is a variable we let x' be a variable. The notion of alphabetical order is understood to be suitably redefined, with the relative order of the old variables being unchanged. Generally speaking, our notation in arithmetizing syntax will correspond to the notation we have been using, following [Sh], for metamathematical discussion, and we will often use primed variables to suggest sequences of expressions, but formally all variables remain on an equal footing.

We will introduce predicate symbols arithmetizing the syntactical notions of variable, predicate symbol, formula, etc. There should be no confusion between the formal use of these predicate symbols and the informal use of the same terminology in talking about expressions.

Our first task is to introduce the various kinds of symbols.

23.1 *Def.* $\mathrm{X}_n = \langle\langle 2, n\rangle\rangle$.

23.2 *Def.* x is a variable $\leftrightarrow \exists n\, x = \mathrm{X}_n$.

23.3 *Def.* $\mathrm{F}_{n,m} = \langle\langle 3, \langle n, m\rangle\rangle\rangle$.

23.4 *Def.* f is a function symbol $\leftrightarrow \exists n \exists m\, f = \mathrm{F}_{n,m}$.

23.5 *Def.* $\mathrm{P}_{n,m} = \langle\langle 4, \langle n, m\rangle\rangle\rangle$.

23.6 *Def.* p is a predicate symbol $\leftrightarrow \exists n \exists m\, p = \mathrm{P}_{n,m}$.

23.7 *Def.* $\bar{\neg} = \langle\langle 5, 0\rangle\rangle$.

23.8 *Def.* $\bar{\vee} = \langle\langle 5, 1\rangle\rangle$.

23.9 *Def.* $\bar{\exists} = \langle\langle 5, 2\rangle\rangle$.

23.10 *Def.* $\bar{\equiv} = \mathrm{P}_{2,0}$.

23.11 *Def.* u is a symbol $\leftrightarrow$ u is a variable $\vee$ u is a function symbol $\vee$ u is a predicate symbol $\vee\ u = \bar{\neg} \vee u = \bar{\vee} \vee u = \bar{\exists}$.

We will tacitly use the theorem that these disjuncts are mutually exclusive.

23.12 *Def.* u is a logical symbol $\leftrightarrow$ u is a variable $\vee\ u = \bar{\equiv} \vee u = \bar{\neg} \vee u = \bar{\vee} \vee u = \bar{\exists}$.

23.13 *Def.* u is a nonlogical symbol $\leftrightarrow$ u is a symbol & $\neg$(u is a logical symbol).

23.14 *Def.* e is a constant $\leftrightarrow \exists m\, e = \mathrm{F}_{0,m}$.

The index of a symbol tells how many arguments it takes:

23.15 *Def.* $\mathrm{Index}\, u = n \leftrightarrow \exists m\, u = \mathrm{F}_{n,m} \vee \exists m\, u = \mathrm{P}_{n,m} \vee (u = \bar{\neg}\ \&\ n = 1) \vee (u = \bar{\vee}\ \&\ n = 2) \vee (u = \bar{\exists}\ \&\ n = 2)$, otherwise $n = 0$.

23.16 *Def.* u is an expression $\leftrightarrow$ u is a sequence & $\forall i(1 \le i \le \mathrm{Ln}\, u \rightarrow \langle u(i)\rangle$ is a symbol).

23.17 *Thm.* u is a symbol $\leftrightarrow$ u is an expression & $\mathrm{Ln}\, u = 1$.

Proof. Suppose u is a symbol. By inspection of the defining axioms, u is a sequence and $\mathrm{Ln}\, u = 1$, so that $u = \langle u(1)\rangle$ and u is an expression. Thus *lhs* (23.17) $\rightarrow$ *rhs* (23.17). The converse is clear. □

Next we want to arithmetize the notion of a term. The usual definition is a generalized inductive definition: an expression is a term in case it is a member of the smallest set of expressions containing the variables and closed under the application of ν-ary function symbols to ν-tuples from the set, for all ν. We can relativize this definition to the set of all expressions occurring in a given expression, because to see whether an expression is a term it is not necessary to look beyond the expressions occurring in it. This is a pattern that we will follow on several occasions.

23.18 *Def.* $\text{Terms}\, u = \text{Min}\, s\,(s \leq \text{Occ}\, u \;\&\; (1) \;\&\; (2))$, *where*

1. $\forall x(x \in \text{Occ}\, u \;\&\; x$ is a variable $\rightarrow x \in s)$,
2. $\forall f \forall v'(f$ is a function symbol $\&$ v' is a sequence of expressions $\&$ $\text{Index}\, f = \text{Ln}\, v'$ $\&$ $\text{Ran}\, v' \subseteq s$ $\&$ $f * v'^{*} \in \text{Occ}\, u \rightarrow f * v'^{*} \in s)$.

23.19 *Def.* u is a term $\leftrightarrow u \in \text{Terms}\, u$.

23.20 *Thm.* v occurs in $u \rightarrow \text{Terms}\, v \subseteq \text{Terms}\, u$.

Proof. Suppopse v occurs in u and let $s_1 = \text{Terms}\, v \cap \text{Terms}\, u$. Then it follows that $scope_{\text{Min}}\,(23.18)_{su}[s_1 v]$, so $\text{Terms}\, v \leq s_1 \leq \text{Terms}\, v$. Thus (23.20).

23.21 *Thm.* a is a term $\rightarrow \forall i(1 \leq i \leq \text{Ln}\, a \rightarrow \langle a(i) \rangle$ is a variable $\vee$ $\langle a(i) \rangle$ is a function symbol).

Proof. Suppose a is a term and let

$$s_1 = \{u \in \text{Terms}\, a : con\,(23.21)_a[u]\}.$$

Then $scope_{\text{Min}}\,(23.18)_{su}[s_1 a]$, so $s_1 = \text{Terms}\, a$. Thus (23.21).

23.22 *Thm.* a is a term $\leftrightarrow a$ is a variable $\vee$ $\exists f \exists a'(f$ is a function symbol $\&$ a' is a sequence of terms $\&$ $\text{Index}\, f = \text{Ln}\, a'$ $\&$ $a = f * a'^{*})$.

Proof. Suppose a is a term and let

$$s_1 = \{u \in \text{Terms}\, a : rhs\,(23.22)_a[u]\}.$$

Then $scope_{\text{Min}}\,(23.18)_{su}[s_1 a]$, and consequently we have $s_1 = \text{Terms}\, a$. Thus $lhs\,(23.22) \rightarrow rhs\,(23.22)$. Conversely, suppose $rhs\,(23.22)$. Clearly a is a variable $\rightarrow a$ is a term, so suppose $scope_{\exists a'}\,(23.22)$. By (23.20), $\text{Ran}\, a' \subseteq \text{Terms}\, a$. Thus a is a term, and thus (23.22).

23.23 *Thm.* $\text{Terms}\, u = \{a \in \text{Occ}\, u : a$ is a term$\}$.

Proof. Let $s = \{a \in \text{Occ}\, u : a$ is a term$\}$. Then $scope_{\text{Min}}\,(23.18)$, so $\text{Terms}\, u \subseteq s$. Suppose $a \in s$. Then $a \in \text{Terms}\, a$, so $a \in \text{Terms}\, u$ by (23.20). Thus $s \subseteq \text{Terms}\, u$, and so (23.23).

23.24 *Thm.* e is a constant $\rightarrow$ e is a term.

Proof. Suppose e is a constant. Then $scope_{\exists a'}(23.22)_{fa'}[e0]$, and thus (23.24).

23.25 *Def.* A is an atomic formula $\leftrightarrow$ $\exists p \exists a'(p$ is a predicate symbol & a' is a sequence of terms & $\text{Index}\, p = \text{Ln}\, a'$ & $A = p * a'^{*})$.

23.26 *Def.* $\text{Formulas}\, u = \text{Min}\, s(s \leq \text{Occ}\, u$ & (1) & (2) & (3) & (4)), *where*

1. $\forall A(A \in \text{Occ}\, u$ & A is an atomic formula $\rightarrow A \in s)$,
2. $\forall A(A \in s$ & $\bar{\neg} * A \in \text{Occ}\, u \rightarrow \bar{\neg} * A \in s)$,
3. $\forall A \forall B(A \in s$ & $B \in s$ & $\bar{\vee} * A * B \in \text{Occ}\, u \rightarrow \bar{\vee} * A * B \in s)$,
4. $\forall A \forall x(A \in s$ & x is a variable & $\bar{\exists} * x * A \in \text{Occ}\, u \rightarrow \bar{\exists} * x * A \in s)$.

23.27 *Def.* u is a formula $\leftrightarrow$ $u \in \text{Formulas}\, u$.

23.28 *Thm.* v occurs in u $\rightarrow$ $\text{Formulas}\, v \subseteq \text{Formulas}\, u$.

Proof. Suppose v occurs in u and let $s_1 = \text{Formulas}\, v \cap \text{Formulas}\, u$. Then $scope_{\text{Min}}(23.26)_{su}[s_1 v]$, so $\text{Formulas}\, v \leq s_1 \leq \text{Formulas}\, v$. Thus (23.28).

23.29 *Thm.* A is a formula $\rightarrow$ A is an expression.

Proof. Suppose A is a formula and let

$$s_1 = \{u \in \text{Formulas}\, A : con\,(23.29)_A[u]\}.$$

Then $scope_{\text{Min}}(23.26)_{su}[s_1 A]$, so $s_1 = \text{Formulas}\, A$. Thus (23.29).

23.30 *Thm.* A is a formula $\rightarrow$ A is an atomic formula $\vee$ $\exists B(B$ is a formula & $A = \bar{\neg} * B)$ $\vee$ $\exists B \exists C(B$ and C are formulas & $A = \bar{\vee} * B * C)$ $\vee$ $\exists x \exists B(x$ is a variable & B is a formula & $A = \bar{\exists} * x * B)$.

Proof. Suppose A is a formula and let

$$s_1 = \{u \in \text{Formulas}\, A : rhs\,(23.20)_A[u]\}.$$

Then $scope_{\text{Min}}(23.26)_{su}[s_1 A]$, so $s_1 = \text{Formulas}\, A$. Thus $lhs\,(23.30) \rightarrow rhs\,(23.30)$. Conversely, suppose $rhs\,(23.30)$. Clearly A is an atomic formula $\rightarrow$ A is a formula, so suppose B and C are formulas & x is a variable & $(A = \bar{\neg} * B \vee A = \bar{\vee} * B * C \vee A = \bar{\exists} * x * B)$. By (23.28), A is a formula. Thus A is a formula, and thus (23.30).

23.31 *Thm.* Formulas $u = \{A \in \text{Occ}\, u : A \text{ is a formula}\}$.

Proof. Let $s = \{A \in \text{Occ}\, u : A \text{ is a formula}\}$. Then $scope_{\text{Min}}$ (23.26), so Formulas $u \subseteq s$. Suppose $A \in s$. Then $A \in$ Formulas A, so $A \in$ Formulas u by (23.28). Thus $s \subseteq$ Formulas u, and so (23.31).

23.32 *Def.* u is a designator $\leftrightarrow$ u is a term $\vee$ u is a formula.

23.33 *Thm.* u is a designator $\rightarrow \exists u'(u'$ is a sequence of designators & Index $\langle u(1)\rangle = \text{Ln}\, u'$ & $u = \langle u(1)\rangle * u'^{*})$.

Proof. By (23.22) and (23.30). □

We want to prove the uniqueness of u'; this is the formation theorem.

23.34 *Def.* u_1 and u_2 are compatible $\leftrightarrow$ u_1 and u_2 are expressions & $\exists v(v$ is an expression & $(u_2 = u_1 * v \vee u_1 = u_2 * v))$.

23.35 *Thm.* u_1 and u_2 are compatible $\leftrightarrow$ u_1 and u_2 are expressions & $\forall i(1 \leq i \leq \text{Ln}\, u_1$ & $1 \leq i \leq \text{Ln}\, u_2 \rightarrow u_1(i) = u_2(i))$.

Proof. Clearly *lhs* (23.35) $\rightarrow$ *rhs* (23.25). Suppose 1: *rhs* (23.35) & $\text{Ln}\, u_1 \leq \text{Ln}\, u_2$, and let $v = u_2[\text{Ln}\, u_1 + 1, \text{Ln}\, u_2]$. Then $u_2 = u_1 * v$, and thus (1) $\rightarrow$ *lhs* (23.35). Therefore *rhs* (23.35) & $\text{Ln}\, u_2 \leq \text{Ln}\, u_1 \rightarrow$ *lhs* (23.35), and so (23.35).

23.36 *Thm.* $u_1 * v_1$ and $u_2 * v_2$ are compatible $\rightarrow$ u_1 and u_2 are compatible.

Proof. By (23.35).

23.37 *Thm.* $u * v_1$ and $u * v_2$ are compatible $\rightarrow$ v_1 and v_2 are compatible.

Proof. By (23.35).

23.38 *Thm.* u' is a sequence & $\forall i(1 \leq i \leq \text{Ln}\, u' \rightarrow u'(i) \neq 0) \rightarrow \text{Ln}\, u' \leq (\sum u')(\text{Ln}\, u')$.

Proof. Suppose $\exists u' \neg$(23.38). By BLNP there exists a minimal such u'. Clearly $u' \neq 0$. Then $\text{Ln Chop}\, u' \leq (\sum \text{Chop}\, u')(\text{Ln Chop}\, u')$, so *con* (23.38), a contradiction, and thus (23.38).

23.39 *Thm.* u' and v' are sequences of designators & $\text{Ln}\, u' = \text{Ln}\, v'$ & u'^{*} and v'^{*} are compatible $\rightarrow u' = v'$.

(See Lemma 1 of [Sh,§2.4].) *Proof.* Suppose (23.39). Write α for

$$\neg(23.39)_{u'v'}[u'_1 v'_1] \ \& \ u'^*_1 \text{ occurs in } u'^* \ \& \\ v'^*_1 \text{ occurs in } v'^* \ \& \ m = \mathrm{Ln}\, u'^*_1.$$

Observe that

$$\exists u'_1 \exists v'_1\, \alpha\colon \mathrm{Ln}\, u'_1 \leq \mathrm{Ln}\, u'^*,\ \mathrm{Sup}\, u'_1 \leq u'^*, \\ \mathrm{Ln}\, v'_1 \leq \mathrm{Ln}\, v'^*,\ \mathrm{Sup}\, v'_1 \leq v'^*$$

by (23.28) and (19.31). We have $\exists m \exists u'_1 \exists v'_1\, \alpha$ (let $u'_1 = u'$, let $v'_1 = v'$, and let $m = \mathrm{Ln}\, u'^*$), so by BLNP there exists m such that $\min_m \exists u'_1 \exists v'_1\, \alpha$, so of course there exist u'_1 and v'_1 such that α. Let $n = \mathrm{Ln}\, u'_1$, so that also $n = \mathrm{Ln}\, v'_1$. Let $u'_2 = \mathrm{Chop}\, u'_1$ and let $v'_2 = \mathrm{Chop}\, v'_1$. Clearly $n \neq 0$. We have $u'^*_1 = u'^*_2 * u'_1(n)$ and $v'^*_1 = v'^*_2 * v'_1(n)$. Since u'^*_1 and v'^*_1 are compatible, u'^*_2 and v'^*_2 are compatible by (23.36). Since $n \neq 0$ we have $\mathrm{Ln}\, u'^*_2 < m$, so by the minimality assumption we have $u'_2 = v'_2$ and consequently $u'^*_2 = v'^*_2$. By (23.37), $u'_1(n)$ and $v'_1(n)$ are compatible. Let $u = u'_1(n)$ and let $v = v'_1(n)$. By (23.33), there exist w' and z' such that w' and z' are sequences of designators & $\mathrm{Index}\,\langle u(1)\rangle = \mathrm{Ln}\, w'$ & $\mathrm{Index}\,\langle v(1)\rangle = \mathrm{Ln}\, z'$ & $u = \langle u(1)\rangle * w'^*$ & $v = \langle v(1)\rangle * z'^*$. By (23.36), $\langle u(1)\rangle$ and $\langle v(1)\rangle$ are compatible, so that $u(1) = v(1)$. By (23.37), w'^* and z'^* are compatible. Also, w'^* occurs in u'^* and z'^* occurs in v'^*, and $\mathrm{Ln}\, w'^* < m$. By the minimality assumption, $w' = z'$ and so $u = v$; that is, $u'_1(n) = v'_1(n)$. We have already seen that $u'_2 = v'_2$; that is, $\mathrm{Chop}\, u'_1 = \mathrm{Chop}\, u'_2$. Therefore $u'_1 = v'_1$, a contradiction, and thus (23.39).

23.40 *Thm.* (*formation theorem*) u is a designator $\to \exists! u'(u'$ is a sequence of designators & $\mathrm{Index}\,\langle u(1)\rangle = \mathrm{Ln}\, u'$ & $u = \langle u(1)\rangle * u'^*)$.

Proof. We have the existence by (23.33). Suppose u is a designator & u'_1 and u'_2 are sequences of designators & $\mathrm{Index}\,\langle u(1)\rangle = \mathrm{Ln}\, u'_1 = \mathrm{Ln}\, u'_2$ & $u = \langle u(1)\rangle * u'^*_1 = \langle u(1)\rangle * u'^*_2$. By (23.37), u'^*_1 and u'^*_2 are compatible, so $u'_1 = u'_2$ by (23.39). Thus (23.39).

23.41 *Def.* $\mathrm{Arg}\, u = u' \leftrightarrow u$ is a designator & u' is a sequence of designators & $\mathrm{Index}\,\langle u(1)\rangle = \mathrm{Ln}\, u'$ & $u = \langle u(1)\rangle * u'^*$, otherwise $u' = 1$.

The uniqueness condition holds by (23.40), and we have $\exists u'\, rhs\,(23.41)$: $\mathrm{Ln}\, u' \leq \mathrm{Ln}\, u$, $\mathrm{Sup}\, u' \leq u$.

23.42 *Thm.* u and v are designators & u and v are compatible $\rightarrow u = v$.

Proof. Suppose *hyp* (23.42). Then $\langle u \rangle$ and $\langle v \rangle$ are sequences of designators, $\mathrm{Ln}\,\langle u \rangle = \mathrm{Ln}\,\langle v \rangle = 1$, $\langle u \rangle^* = u$, and $\langle v \rangle^* = v$, so by (23.39) we have $\langle u \rangle = \langle v \rangle$. Hence $u = v$, and thus (23.42).

23.43 *Thm.* u is a designator & $1 \leq i \leq \mathrm{Ln}\,u \rightarrow \exists j(i \leq j \leq \mathrm{Ln}\,u$ & $u[i,j]$ is a designator).

(See Lemma 2 of [Sh,§2.4].) *Proof.* Suppose $\exists u \exists i \neg$(23.43). By BLNP there exist minimal such u and i. Suppose $i = 1$ and let $j = \mathrm{Ln}\,u$. Then *con* (23.43), a contradiction, and thus $i \neq 1$. Let $v' = \mathrm{Arg}\,u$ and let $v_1' = \langle\langle u(1) \rangle\rangle * v'$. Recall (19.28) and (19.29), and let $k = \mathrm{Loc}_1(i, v_1')$ and let $i_1 = \mathrm{Loc}_2(i, v_1')$. Then $v'(k) < u$ and $i_1 \leq i$, so by the minimality assumption there exists j_1 such that $i_1 \leq j_1 \leq v'(k)$ & $v'(k)[i_1, j_1]$ is a designator. Let $j = i + j_1 - i_1$. Then $u[i,j] = v'(k)[i_1, j_1]$ and so *con* (23.43), a contradiction, and thus (23.43).

23.44 *Thm.* (*occurrence theorem*) u is a designator & $1 \leq i \leq j \leq \mathrm{Ln}\,u$ & $u[i,j]$ is a designator $\rightarrow u[i,j] = u \vee \exists k \exists i_1 \exists j_1(1 \leq k \leq \mathrm{Ln}\,\mathrm{Arg}\,u$ & $1 \leq i_1 \leq j_1 \leq \mathrm{Ln}\,((\mathrm{Arg}\,u)(k))$ & $u[i,j] = (\mathrm{Arg}\,u)(k)[i_1, j_1])$.

Proof. Suppose *hyp* (23.44). Suppose $i = 1$. Then $u[i,j]$ and u are compatible, so $u[i,j] = u$ by (23.42). Thus $i = 1 \rightarrow$ *con* (23.44), so suppose $i \neq 1$. Let $u' = \mathrm{Arg}\,u$, let $v' = \langle\langle u(1) \rangle\rangle * u'$, let $k = \mathrm{Loc}_1(i, v')$, and let $i_1 = \mathrm{Loc}_2(i, v)$. By (23.43) there exists j_1 such that $i_1 \leq j_1 \leq \mathrm{Ln}\,u'(k)$ & $u'(k)[i_1, j_1]$ is a designator. Then $u[i,j]$ and $u'(k)[i_1, j_1]$ are compatible, so $u[i,j] = u'(k)[i_1, j_1]$ by (23.42). Thus $i \neq 1 \rightarrow$ *con* (23, 44), and thus (23.44).

23.45 *Def.* i is a bound occurrence of x in $A \leftrightarrow x$ is a variable & A is a formula & $\langle A(i) \rangle = x$ & $\exists B \exists j(B$ is a formula & $1 \leq j < i \leq j + 2 + \mathrm{Ln}\,B \leq \mathrm{Ln}\,A$ & $A[j, j + 2 + \mathrm{Ln}\,B] = \bar{\exists} * x * B)$.

23.46 *Def.* i is a free occurrence of x in $u \leftrightarrow x$ is a variable & u is a deignator & $\langle u(i) \rangle = x$ & $\neg(i$ is a bound occurrence of x in $u)$.

23.47 *Def.* x is bound in $A \leftrightarrow \exists i(i$ is a bound occurrence of x in $A)$.

23.48 *Def.* x is free in $u \leftrightarrow \exists i(i$ is a free occurrence of x in $u)$.

When we substitute terms for variables in a formula, we do so only for free occurrences of the variables, so we introduce the following defining axioms:

23.49 *Def.* $\mathrm{Subfr}(u, x, a) = \{\langle i, v\rangle : i \in \mathrm{Dom}\, u\ \&\ (\neg(i \text{ is a free occurrence of } x \text{ in } u) \to v = \langle u(i)\rangle)\ \&\ (i \text{ is a free occurrence of } x \text{ in } u \to v = a)\}^*$.

23.50 *Def.* $\mathrm{Ssubfr}(u, x', a') = \{\langle i, v\rangle : i \in \mathrm{Dom}\, u\ \&\ (\neg\exists j(1 \le j \le \mathrm{Ln}\, x'\ \&\ i \text{ is a free occurrence of } x'(j) \text{ in } u\ ) \to v = \langle u(i)\rangle)\ \&\ \forall j(\mathrm{min}_j(1 \le j \le \mathrm{Ln}\, x'\ \&\ i \text{ is a free occurrence of } x'(j) \text{ in } u) \to v = a'(j))\}^*$.

23.51 *Def.* $u_x\lceil a\rceil = v \leftrightarrow (1) \vee (2)$, otherwise $v = 1$, *where*

1. u is a designator & x is a variable & a is a term & $v = \mathrm{Subfr}(u, x, a)$,
2. u is a designator & x is an injective sequence of variables & a is a sequence of terms & $\mathrm{Ln}\, x = \mathrm{Ln}\, a$ & $v = \mathrm{Ssubfr}(u, x, a)$.

23.52 *Thm.* u is a designator & ((x is a variable & a is a term) $\vee$ (x is an injective sequence of variables & a is a sequence of terms & $\mathrm{Ln}\, x = \mathrm{Ln}\, a$)) $\to$ ($u_x\lceil a\rceil$ is a term $\leftrightarrow$ u is a term) & ($u_x\lceil a\rceil$ is a formula $\leftrightarrow$ u is a formula).

Proof. Suppose $\exists u \exists x \exists a \neg(23.52)$. By BLNP there exist minimal such u, x, and a. Let $u' = \mathrm{Arg}\, u$ and let

$$u'_1 = \{\langle k, u'(k)_x\lceil a\rceil\rangle : k \in \mathrm{Dom}\, u\}.$$

Then by the minimality assumption we have

$$\forall k(1 \le k \le \mathrm{Ln}\, u' \to \mathit{con}\,(23.52)_u\lceil u'(k)\rceil).$$

Suppose $\langle u(1)\rangle$ is a variable. Then $\mathrm{Index}\,\langle u(1)\rangle = 0$, so $\mathrm{Ln}\, u = 1$ and $u = \langle u(1)\rangle$. But then $u_x\lceil a\rceil$ is a term, so *con* (23.52), a contradiction. Thus $\neg(\langle u(1)\rangle$ is a variable). Suppose $\langle u(1)\rangle \ne \bar{\exists}$. Then $u_x\lceil a\rceil = \langle u(1)\rangle * u_1'^*$ and so *con* (23.52), a contradiction. Thus $\langle u(1)\rangle = \bar{\exists}$, so there exist y and B such that y is a variable & B is a formula & $u = \bar{\exists} * y * B$, so that $u'^* = y * B$. Suppose 1: (x is a variable & $x \ne y$) $\vee$ (x is a sequence of variables & $y \notin \mathrm{Ran}\, x$). Then again $u_x\lceil a\rceil = \langle u(1)\rangle * u_1'^*$ and so *con* (23.52), a contradiction. Thus $\neg(1)$. Suppose 2: $x = y$. Then $u_x\lceil a\rceil = u$, so *con* (23.52), a contradiction. Thus $\neg(2)$. Therefore x is a sequence of variables & $y \in \mathrm{Ran}\, x$. Then there exists j such that

$1 \le j \le \mathrm{Ln}\, x$ & $x(j) = y$. Let $x_1 = x[1, j-1] * x[j+1, \mathrm{Ln}\, x]$ and let $a_1 = a[1, j-i] * a[j+1, \mathrm{Ln}\, a]$. By (19.38), $x_1 < x$ and $a_1 < a$, so by the minimality assumption we have *con* $(23.52)_{uxa}[Bx_1a_1]$. Since x is injective, $y \notin \mathrm{Ran}\, x_1$. Hence $u_x\lceil a \rceil = \bar{\exists} * y * B_{x_1}\lceil a_1 \rceil$, so *con* (23.52), a contradiction. Thus (23.52).

23.53 *Def.* a is substitutable for x in $A \leftrightarrow a$ is a term & x is a variable & A is a formula & $\forall y \forall B \forall i \forall j \forall k(y$ is a variable & y occurs in a & B is a formula & $1 \le i \le j \le k \le \mathrm{Ln}\, A$ & $A[i,k] = \bar{\exists} * y * B \dashrightarrow \neg(j$ is a free occurrence of x in $A))$.

23.54 *Def.* a' is simultaneously substitutable for x' in $A \leftrightarrow a'$ is a sequence of terms & x' is an injective sequence of variables & $\mathrm{Ln}\, x' = \mathrm{Ln}\, a'$ & $\forall i(1 \le i \le \mathrm{Ln}\, x' \to a'(i)$ is substitutable for $x'(i)$ in $A)$.

Now we introduce some function symbols that enable us to write formulas more compactly.

23.55 *Def.* $\tilde{\neg}A = C \leftrightarrow A$ is a formula & $C = \bar{\neg} * A$, otherwise $C = 1$.

23.56 *Def.* $A \tilde{\vee} B = C \leftrightarrow A$ and B are formulas & $C = \bar{\vee} * A * B$, otherwise $C = 1$.

23.57 *Def.* $\tilde{\exists} x A = C \leftrightarrow x$ is a variable & A is a formula & $C = \bar{\exists} * x * A$, otherwise $C = 1$.

23.58 *Def.* $\tilde{\forall} x A = \tilde{\neg} \tilde{\exists} x \tilde{\neg} A$.

23.59 *Def.* $A \tilde{\to} B = \tilde{\neg} A \tilde{\vee} B$.

23.60 *Def.* $A \tilde{\&} B = \tilde{\neg}(A \tilde{\to} \tilde{\neg} B)$.

23.61 *Def.* $A \tilde{\leftrightarrow} B = (A \tilde{\to} B) \tilde{\&} (B \tilde{\to} A)$.

23.62 *Def.* $a \tilde{=} b = C \leftrightarrow a$ and b are terms & $C = \bar{=} * a * b$, otherwise $C = 1$.

23.63 *Def.* $a \tilde{\neq} b = \tilde{\neg}(a \tilde{=} b)$.

We need to express iterated disjunctions, implications, and conjunctions, associated from right to left.

23.64 *Def.* $\mathrm{Disj}\, A' = \{\langle i, u \rangle : i \in \mathrm{Dom}\, A'$ & $(i = \mathrm{Ln}\, A' \to u = A'(\mathrm{Ln}\, A'))$ & $(i < \mathrm{Ln}\, A' \to u = \bar{\vee} * A'(i))\}^*$.

23.65 *Def.* $\operatorname{Impl} A' = \{\langle i, u\rangle : i \in \operatorname{Dom} A' \ \& \ (i = \operatorname{Ln} A' \to u = A'(\operatorname{Ln} A')) \ \& \ (i < \operatorname{Ln} A' \to u = \tilde{\vee} * \tilde{\neg} * A'(i))\}^*$.

23.66 *Def.* $\operatorname{Conj} A' = \{\langle i, u\rangle : i \in \operatorname{Dom} A' \ \& \ (i = \operatorname{Ln} A' \to u = A'(\operatorname{Ln} A')) \ \& \ (i < \operatorname{Ln} A' \to u = \tilde{\neg} * \tilde{\vee} * \tilde{\neg} * A'(i) * \tilde{\neg})\}^*$.

23.67 *Thm.* A' is a sequence of formulas & $A' \neq 0$ & $A = A'(1)$ & $B' = A'[2, \operatorname{Ln} A'] \to \operatorname{Disj} A'$, $\operatorname{Impl} A'$, and $\operatorname{Conj} A'$ are formulas & $(\operatorname{Ln} A' = 1 \to \operatorname{Disj} A' = \operatorname{Impl} A' = \operatorname{Conj} A' = A)$ & $(\operatorname{Ln} A' \neq 1 \to \operatorname{Disj} A' = A \tilde{\vee} \operatorname{Disj} B'$ & $\operatorname{Impl} A' = A \tilde{\to} \operatorname{Impl} B'$ & $\operatorname{Conj} A' = A \tilde{\&} \operatorname{Conj} B')$.

Proof. Suppose $\exists A' \neg (23.67)$. By BLNP there exists a minimal such A'. Suppose $\operatorname{Ln} A' = 1$. Then $\operatorname{Disj} A' = \operatorname{Impl} A' = \operatorname{Conj} A' = \langle A\rangle^*$, but A is a sequence and so $\langle A\rangle^* = A$, a contradiction. Thus $\operatorname{Ln} A' \neq 1$. Then $\operatorname{Disj} A' = \tilde{\vee} * A * \operatorname{Disj} B' = A \tilde{\vee} \operatorname{Disj} B'$, and $\operatorname{Impl} A' = \tilde{\vee} * \tilde{\neg} * A * \operatorname{Impl} B' = \tilde{\neg} A \tilde{\vee} \operatorname{Impl} B' = A \tilde{\to} \operatorname{Impl} B'$, and $\operatorname{Conj} A = \tilde{\neg} * \tilde{\vee} * \tilde{\neg} * A * \tilde{\neg} * \operatorname{Conj} B' = \tilde{\neg}(A \tilde{\to} \tilde{\neg} \operatorname{Conj} B') = A \tilde{\&} \operatorname{Conj} B'$. This is a contradiction, and thus (23.67). □

Using this theorem, we see that if a is $\langle A_1\rangle * \langle A_2\rangle * \cdots * \langle A_\nu\rangle$, then we have: a is a sequence of formulas $\to \operatorname{Disj} a = A_1 \tilde{\vee} A_2 \tilde{\vee} \cdots \tilde{\vee} A_\nu$ & $\operatorname{Impl} a = A_1 \tilde{\to} A_2 \tilde{\to} \cdots \tilde{\to} A_\nu$ & $\operatorname{Conj} a = A_1 \tilde{\&} A_2 \tilde{\&} \cdots \tilde{\&} A_\nu$. Unless the contrary is stated, a binary operation is associated from right to left when restoring parentheses; for example, $A \tilde{\to} B \tilde{\to} C$ is $A \tilde{\to} (B \tilde{\to} C)$.

23.68 *Def.* u is variable-free $\leftrightarrow$ u is an expression & $\forall i(1 \leq i \leq \operatorname{Ln} u \to \neg(\langle u(i)\rangle$ is a variable$))$.

23.69 *Def.* A is a closed formula $\leftrightarrow$ A is a formula & $\forall i(1 \leq i \leq \operatorname{Ln} A \to \neg(i$ is a free occurrence of $\langle A(i)\rangle$ in $A))$.

Chapter 24

Proofs

We give a predicative arithmetization of the predicate calculus. We modify the treatment in [Sh,§2.6] by adopting tautological consequence as a rule of inference; see the conclusion of [Sh,§3.1].

24.1 *Def.* B is a substitution axiom $\leftrightarrow \exists A \exists x \exists a(a$ is substitutable for x in A & $B = A_x[a] \tilde{\rightarrow} \tilde{\exists} x A)$.

24.2 *Def.* B is an identity axiom $\leftrightarrow \exists x(x$ is a variable & $B = x \tilde{=} x)$.

24.3 *Def.* $\text{Equals}(x', y') = \{\langle i, x'(i) \tilde{=} y'(i) \rangle : i \in \text{Dom}\, x'\}$.

24.4 *Def.* B is an equality axiom $\leftrightarrow \exists x' \exists y'(x'$ and y' are sequences of variables & $\text{Ln}\, x' = \text{Ln}\, y'$ & $((1) \vee (2)))$, *where*

1. $\exists f(f$ is a function symbol & $\text{Index}\, f = \text{Ln}\, x'$ & $B = \text{Impl}\,(\text{Equals}(x', y') * \langle f {*} x'^{*} \tilde{=} f {*} y'^{*} \rangle))$,
2. $\exists p(p$ is a predicate symbol & $\text{Index}\, p = \text{Ln}\, x'$ & $B = \text{Impl}\,(\text{Equals}(x', y') * \langle p {*} x'^{*} \tilde{\rightarrow} p {*} y'^{*} \rangle))$.

24.5 *Def.* B is a logical axiom $\leftrightarrow$ B is a substitution axiom $\vee$ B is an identity axiom $\vee$ B is an equality axiom.

24.6 *Def.* v is a truth valuation on $u \leftrightarrow \text{Dom}\, v = \text{Formulas}\, u$ & $\text{Ran}\, v \subseteq \{0\} \cup \{1\}$ & $\forall B(\tilde{\neg} B \in \text{Formulas}\, u \rightarrow v(\tilde{\neg} B) = 1 - v(B))$ & $\forall B \forall C(B \tilde{\vee} C \in \text{Formulas}\, u \rightarrow v(B \tilde{\vee} C) = \text{Max}\,(v(B), v(C)))$.

24.7 *Def.* A is a tautology $\leftrightarrow$ A is a formula & $\forall v(v$ is a truth valuation on $A \rightarrow v(A) = 1)$.

We have *rhs* (24.7): $\operatorname{Dom} v \leq \text{Formulas } A, \operatorname{Sup} v \leq 1$.

24.8 *Def.* A is a tautological consequence of $A' \leftrightarrow A$ is a formula & A' is a sequence of formulas & $\operatorname{Impl}(A' * \langle A \rangle)$ is a tautology.

24.9 *Def.* C can be inferred from D by $\exists$-introduction $\leftrightarrow \exists x \exists A \exists B(x$ is a variable & A and B are formulas & $\neg(x$ is free in $B)$ & $D = A \tilde{\rightarrow} B$ & $C = \tilde{\exists} x(A \tilde{\rightarrow} B))$.

24.10 *Def.* l is a language $\leftrightarrow l$ is a set of nonlogical symbols.

24.11 *Def.* $\operatorname{Nls} u = \operatorname{Ran}\{\langle i, v \rangle : i \in \operatorname{Dom} u \ \&\ v = \langle u(i) \rangle$ & v is a nonlogical symbol$\}$.

24.12 *Def.* a is a term of $l \leftrightarrow a$ is a term & $\operatorname{Nls} a \subseteq l$.

24.13 *Def.* A is a formula of $l \leftrightarrow A$ is a formula & $\operatorname{Nls} A \subseteq l$.

24.14 *Def.* t is a theory $\leftrightarrow \exists l \exists s(l$ is a language & s is a set of formulas of l & $t = \langle l, s \rangle)$.

24.15 *Def.* $\operatorname{Lang} t = l \leftrightarrow t$ is a theory & $l = \operatorname{Proj}_1 t$, otherwise $l = 1$.

24.16 *Def.* $\operatorname{Ax} t = s \leftrightarrow t$ is a theory & $s = \operatorname{Proj}_2 t$, otherwise $s = 1$.

The arithmetical nature of logic is revealed by the fact that Lang and Ax determine a theory.

24.17 *Def.* D' is a proof in t of $A \leftrightarrow D'$ is a sequence of formulas of $\operatorname{Lang} t$ & $D' \neq 0$ & $D'(\operatorname{Ln} D') = A$ & $\forall i(1 \leq i \leq \operatorname{Ln} D' \rightarrow A'(i)$ is a logical axiom $\vee A'(i) \in \operatorname{Ax} t \vee A'(i)$ is a tautological consequence of $A'[1, i-1] \vee \exists j(1 \leq j < i$ & $A'(i)$ can be inferred from $A'(j)$ by $\exists$-introduction$))$.

24.18 *Def!* A is a theorem of $t \leftrightarrow \exists D'(D'$ is a proof in t of $A)$.

Throughout this book I have been quibbling about exponential bounds being unsatisfactory, but here is a predicate symbol that is utterly unbounded. This is the fascination of mathematics.

24.19 *Def.* $f_{|s} = \{z \in f : \operatorname{Proj}_1 z \in s\}$.

24.20 *Thm.* A' is a sequence of formulas & $A' \neq 0$ & v is a truth valuation on $\operatorname{Disj} A' \rightarrow (v(\operatorname{Disj} A') = 1 \leftrightarrow \exists i(1 \leq i \leq \operatorname{Ln} A'$ & $v(A'(i)) = 1))$.

Proof. Suppose $\exists A' \exists v \neg(24.20)$. By BLNP there exist minimal such A' and v. Let $A = A'(i)$, let $B' = A'[2, \operatorname{Ln} A']$, and let $v_0 = v_{|\text{Formulas Disj } B'}$.

Clearly $B' \neq 0$, so we have *con* $(24.20)_{A'v}[B'v_0]$. But by (23.67) we have $v(\mathrm{Disj}\, A') = \mathrm{Max}\,(v(A), v(\mathrm{Disj}\, B'))$, so *con* (24.20), a contradiction, and thus (24.20).

24.21 *Thm.* A' is a sequence of formulas & $A' \neq 0$ & v is a truth valuation on $\mathrm{Impl}\, A' \to (v(\mathrm{Impl}\, A') = 1 \leftrightarrow \exists i(1 \leq i < \mathrm{Ln}\, A' \ \& \ v(A'(i)) = 0) \lor v(A'(\mathrm{Ln}\, A')) = 1)$.

Proof. Suppose $\exists A' \exists v \neg(24.21)$. By BLNP there exists minimal such A' and v. Let $A = A'(1)$, let $B = A'[2, \mathrm{Ln}\, A]$, and let $v_0 = v_{|\mathrm{Formulas\,Impl}\, B'}$. Clearly $B' \neq 0$, so we have *con* $(24.21)_{A'v}[B'v_0]$. But by (23.67) we have $v(\mathrm{Impl}\, A') = \mathrm{Max}\,(1 - v(A), v(\mathrm{Impl}\, B'))$, so *con* (24.21), a contradiction, and thus (24.21).

24.22 *Thm.* A' is a sequence of formulas & $A' \neq 0$ & v is a truth valuation on $\mathrm{Conj}\, A' \to (v(\mathrm{Conj}\, A') = 1 \leftrightarrow \forall i(1 \leq i \leq \mathrm{Ln}\, A' \to v(A'(i)) = 1))$.

Proof. Suppose $\exists A' \exists v \neg(24.22)$. By BLNP there exist minimal such A' and v. Let $A = A'(1)$, $B' = A'[2, \mathrm{Ln}\, A']$, and let $v_0 = v_{|\mathrm{Formulas\,Conj}\, B'}$. Clearly $B' \neq 0$, so we have *con* $(24.22)_{A'v}[B'v_0]$. But by (23.67) we have $v(\mathrm{Conj}\, A') = v(A) \cdot v(\mathrm{Conj}\, B')$, and so *con* (24.22), a contradiction, and thus (24.22).

24.23 *Thm.* A' is a sequence of formulas & A is a formula $\to$ A is a tautological consequence of $A' * \langle \mathrm{Impl}\,(A' * \langle A \rangle)\rangle$.

Proof. Suppose *hyp* (24.23) and let $B = \mathrm{Impl}\,(A' * \langle \mathrm{Impl}\,(A' * \langle A \rangle)\rangle * \langle A \rangle)$. Suppose v is a truth valuation on B & $v(B) = 0$. Then, by (24.21), $v(A) = 0$, $\forall i(1 \leq i \leq \mathrm{Ln}\, A' \to v(A'(i)) = 1)$, and $v(\mathrm{Impl}\,(A' * \langle A \rangle)) = 1$. But this contradicts (24.21), and thus B is a tautology. Thus (24.23).

24.24 *Thm.* A' is a sequence of formulas & $\mathrm{Ran}\, A' \subseteq \mathrm{Occ}\, C$ & v is a truth valuation on $C \to \exists v_0(v_0$ is a truth valuation on $\mathrm{Impl}\, A'$ & $\forall i(1 \leq i \leq \mathrm{Ln}\, A' \to v_0(A'(i)) = v(A'(i))))$.

Proof. We have (24.24): $\mathrm{Dom}\, v_0 \leq \mathrm{Formulas\,Impl}\, A'$, $\mathrm{Sup}\, v_0 \leq 1$. Suppose $\exists A' \neg(24.24)$. By BLNP there exists a minimal such A'. Clearly $\mathrm{Ln}\, A' \geq 2$. Let $A = A'(1)$ and let $B' = A'[2, \mathrm{Ln}\, A']$. Then there exists v_1 such that

$$scope_{\exists v_0}\,(24.24)_{v_0 A'}[v_1 B'].$$

By (23.67), $\text{Impl}\, A' = A \,\tilde{\to}\, \text{Impl}\, B'$. Let

$$\begin{aligned} v_0 = \{\langle B, z\rangle : {} & B \in \text{Formulas Impl}\, A' \;\;\& \\ & (B = \text{Impl}\, A' \to z = \text{Max}\,(1 - v(A), v_1(\text{Impl}\, B'))) \;\;\& \\ & (B = \tilde{\neg} A \to z = 1 - v(A)) \;\;\& \\ & (B \in \text{Formulas}\, A \to z = v(B)) \;\;\& \\ & (B \in \text{Formulas Impl}\, B' \to z = v_1(B))\}. \end{aligned}$$

By the occurrence theorem (23.44), $\text{Dom}\, v_0 = \text{Formulas Impl}\, A'$, and we have *con* (24.24), a contradiction. Thus (24.24).

24.25 *Thm.* A is a tautological consequence of A' & B' is a sequence of formulas & $\text{Ran}\, A' \subseteq \text{Ran}\, B' \to A$ is a tautological consequence of B'.

Proof. Suppose *hyp* (24.25) and suppose v is a truth valuation on $\text{Impl}\,(B' * \langle A\rangle)$ & $v(\text{Impl}\,(B' * \langle A\rangle)) = 0$. By (24.21), $v(A) = 0$ and $\forall i(1 \le i \le \text{Ln}\, B' \to v(B'(i)) = 1)$. Therefore $\forall i(1 \le i \le \text{Ln}\, A' \to v(A'(i)) = 1)$. By (24.24) there exists v_0 such that v_0 is a truth valuation on $\text{Impl}\,(A' * \langle A\rangle)$ & $\forall i(1 \le i \le \text{Ln}\, A' \to v_0(A'(i)) = 1)$ & $v_0(A) = 0$. This contradicts (24.21), and thus *con* (24.25). Thus (24.25).

24.26 *Thm.* A is a tautological consequence of A' & $\forall i(1 \le i \le \text{Ln}\, A' \to D''(i)$ is a proof in t of $A'(i))$ & A is a formula of $\text{Lang}\, t \to D''^{*} * \langle A\rangle$ is a proof in t of A.

Proof. By (24.25). □

Using semibounded replacement, we can prove 1: A is a tautological consequence of A' & $\forall i(1 \le i \le \text{Ln}\, A' \to A'(i)$ is a theorem of $t)$ & A is a formula of $\text{Lang}\, t \to A$ is a theorem of t. The proof goes as follows. Suppose *hyp* (1). Using SBRD, let

$$D'' = \{\langle i, D'\rangle : \; i \in \text{Dom}\, A' \;\;\& \;\; \min_{D'}(D' \text{ is a proof in } t \text{ of } A'(i))\}.$$

By (24.26) we have *con* (1), and thus (1). We need semibounded replacement to know that a sequence of theorems has a sequence of proofs!

Chapter 25

Derived rules of inference

The material in [Sh,§§3.2–3.5] is concerned with derived rules of inference. Straightforward arithmetizations of all of these results are theorems of our theory. All of these derived rules of inference can be expressed by bounded function symbols. All of the induction arguments in these sections of [Sh] are bounded, with one exception—and an alternate predicative proof can be given for it. The reader who is willing to accept these conclusions should read on for a few paragraphs, where some notational conventions are introduced, and then skip the remainder of this lengthy chapter.

Sometimes in the course of the proof of a theorem (ξ) we will write: define $\mathrm{y} = \mathrm{a}$. We do this only when y does not occur in a and a is a bounded term. Let $\mathrm{x}_1, \dots, \mathrm{x}_\mu$ be the variables in a, in the following order: first those occurring free in (ξ) in the order of their occurrence in (ξ) and then the remaining ones in the order of their occurrence in a. Then we regard "define $\mathrm{y} = \mathrm{a}$" as an abbreviation for

$$\textit{Def}.\ \mathrm{y}_\xi \mathrm{x}_1 \dots \mathrm{x}_\mu = \mathrm{y} \leftrightarrow \mathrm{y} = \mathrm{a}.$$

This is the defining axiom of a bounded function symbol y_ξ. Within the proof of (ξ) after the introduction of the function symbol y_ξ, we use y as an abbreviation for the term $\mathrm{y}_\xi \mathrm{x}_1 \dots \mathrm{x}_\mu$. This device, especially when iterated, saves us from constantly having to display the arguments of the function symbols. The theorem (ξ) itself may contain function symbols of the form y_ξ; this may seem odd, but it allows us to introduce the defining axioms of the function symbols at the natural point in the exposition.

Observe that if x and z do not occur in A, then

$$\text{MIND}'.\ \text{Min}\,\mathrm{x}\,\mathrm{A} = \mathrm{z} \leftrightarrow \min_{\mathrm{z}} \mathrm{A}_{\mathrm{x}}[\mathrm{z}],\ \text{otherwise } \mathrm{z} = 0$$

is the defining axiom of a function symbol, since the existence and uniqueness conditions hold automatically. It may or may not be a bounded function symbol, and if it is a bounded function symbol then its defining axiom (MIND′) is equivalent to a defining axiom of the form (MIND). If we ever have occasion to use this notation to introduce a function symbol for which no claim is made that it is bounded, a "!" will indicate this, but the intention is to use (MIND′) to save writing obvious bounds explicitly. Whenever a term of the form Min x A occurs, it is understood that the corresponding defining axiom of the form (MIND′) has been adjoined to the theory.

If x occurs in A only in the part $\exists$xB, then we write

define x as in A

for

define x = Min x B.

We write "define $\mathrm{y}_1 = \mathrm{a}_1, \ldots, \mathrm{y}_\nu = \mathrm{a}_\nu$" for "define $\mathrm{y}_1 = \mathrm{a}_1$, ..., and define $\mathrm{y}_\nu = \mathrm{a}_\nu$". We write "define $\mathrm{x}_1, \ldots, \mathrm{x}_\nu$ as in A" for "define x_1 as in A, ..., and define x_ν as in A".

If we have formulas labeled μ for $\lambda \leq \mu \leq \nu$, and terms a_μ for $\lambda \leq \mu \leq \nu$, we write

define $\mathrm{y} = \mathrm{a}_\mu$ in case (μ), for $\lambda \leq \mu \leq \nu$

for

$$\text{define } \mathrm{y} = \text{Min}\,\mathrm{y}(((\lambda) \rightarrow \mathrm{y} = \mathrm{a}_\lambda)\ \&\ \cdots\ \&\ ((\nu) \rightarrow \mathrm{y} = \mathrm{a}_\nu)).$$

25.1 *Def.* D': $A' \tilde{\vdash} A \leftrightarrow D'$ is a proof in $\langle \text{Nls}\, A'^* * A, \text{Ran}\, A' \rangle$ of A.

25.2 *Thm.* D': $A' \tilde{\vdash} A$ & A is a formula of Lang t & $\forall i(1 \leq i \leq \text{Ln}\, A' \rightarrow C''(i)$ is a proof in t of $A'(i)) \rightarrow C''^* * D'$ is a proof in t of A.

Proof. By (24.25). □

Now we give some results on quantifiers, following [Sh,§3.2].

25.3 *Thm.* ($\forall$-*introduction*) A and B are formulas & x is a variable & $\neg(x$ is free in $A) \to D'_{25.3}(A, B, x)$: $\langle A \tilde{\to} B\rangle \tilde{\vdash} A \tilde{\to} \tilde{\forall} x B$.

Proof. Define $D' = \langle A \tilde{\to} B\rangle * \langle \tilde{\neg} B \tilde{\to} \tilde{\neg} A\rangle * \langle \tilde{\exists} x \tilde{\neg} B \tilde{\to} \tilde{\neg} A\rangle * \langle A \tilde{\to} \tilde{\forall} x B\rangle$.

25.4 *Thm.* (*generalization*) A is a formula & x is a variable $\to$ $D'_{25.4}(A, x)$: $\langle A\rangle \tilde{\vdash} \tilde{\forall} x A$.

Proof. Define $D' = \langle A\rangle * \langle \tilde{\neg} A \tilde{\to} \tilde{\forall} x A\rangle * \langle \tilde{\exists} x \tilde{\neg} A \tilde{\to} \tilde{\forall} x A\rangle * \langle \tilde{\forall} x A\rangle$.

25.5 *Thm.* a is substitutable for x in $A \to D'_{25.5}(a, x, A)$: $\langle A\rangle \tilde{\vdash} A_x\lceil a\rceil$.

Proof. Define $D' = \langle A\rangle * D'_{25.4}(A, x) * \langle \tilde{\forall} x A\rangle * \langle \tilde{\neg} A_x\lceil a\rceil \tilde{\to} \tilde{\exists} x \tilde{\neg} A\rangle * \langle \tilde{\forall} x A \tilde{\to} A_x\lceil a\rceil\rangle * \langle A_x\lceil a\rceil\rangle$. □

Notice that $\langle A\rangle$ and $\langle \tilde{\forall} x A\rangle$ are already contained in $D'_{25.4}(A, x)$, but this redundancy makes the proof somewhat easier to read.

25.6 *Thm.* (*substitution rule*) a' is simultaneously substitutable for x' in A $\to D'_{25.6}(a', x'A)$: $\langle A\rangle \tilde{\vdash} A_{x'}\lceil a'\rceil$.

Proof. Define

$$
\begin{aligned}
&y' = \{\langle i, X_{a'+x'+A+i}\rangle : i \in \operatorname{Dom} x'\},\\
&A'_0 = \{\langle i, A_{x'[1,i-1]}\lceil y'[1, i-1]\rceil\rangle : i \in \operatorname{Dom} x' \cup \{\operatorname{Ln} x' + 1\}\},\\
&B'_0 = \{\langle i, D'_{25.5}(y'(i), x'(i), A'_0(i))\rangle : i \in \operatorname{Dom} x'\}^*,\\
&B = A_{x'}\lceil y'\rceil,\\
&A'_1 = \{\langle i, B_{y'[1,i-1]}\lceil a'[1, i-1]\rceil\rangle : i \in \operatorname{Dom} y' \cup \{\operatorname{Ln} y' + 1\}\},\\
&B'_1 = \{\langle i, D'_{25.5}(a'(i), y'(i), A'_1(i))\rangle : i \in \operatorname{Dom} y'\}^*,\\
&D' = B'_0 * B'_1.
\end{aligned}
$$

Suppose *hyp* (25.6). We have $\forall i(1 \le i \le \operatorname{Ln} x' \to y'(i)$ is substitutable for $x'(i)$ in $A'_0(i)$ & $a'(i)$ is substitutable for $y'(i)$ in $A'_1(i))$, so by (25.5) we have B'_0: $\langle A\rangle \tilde{\vdash} B$ and B'_1: $\langle B\rangle \tilde{\vdash} B_{y'}\lceil a'\rceil$. But $B_{y'}\lceil a'\rceil = A_{x'}\lceil a'\rceil$, so *con* (25.6) and thus (25.6).

25.7 *Def.* A_0 is an instance of $A \leftrightarrow \exists x' \exists a'(a'$ is simultaneously substitutable for x' in A & $A_0 = A_{x'}\lceil a'\rceil)$.

25.8 *Thm.* A_0 is an instance of $A \to D'_{25.8}(A_0, A)$: $\langle A\rangle \tilde{\vdash} A_0$.

Proof. Define x' and a' as in (25.7), and define $D' = D'_{25.6}(a', x', A)$.

25.9 *Def.* Exist $x' = \{\langle i, \exists * x'(i)\rangle : i \in \operatorname{Dom} x'\}^*$.

25.10 *Def.* $\text{All}\, x' = \{\langle i, \tilde{\neg} * \tilde{\exists} * x'(i) * \tilde{\neg} \rangle : i \in \text{Dom}\, x'\}^*$.

25.11 *Thm.* x' is a sequence of variables & A is a formula $\rightarrow$ $\text{Exist}\, x' * A$ and $\text{All}\, x' * A$ are formulas.

Proof. Suppose $\exists x' \neg (25.11)$. By BLNP there exists a minimal such x'. Clearly $\text{Ln}\, x' \geq 2$. Let $x'_1 = x'[2, \text{Ln}\, x']$. Then $\text{Exist}\, x'_1 * A$ and $\text{All}\, x'_1 * A$ are formulas, but $\text{Exist}\, x' * A = \tilde{\exists} x'(1)(\text{Exist}\, x'_1 * A)$ and $\text{All}\, x' * A = \tilde{\forall} x'(1)(\text{All}\, x'_1 * A)$, so *con* (25.11), a contradiction. Thus (25.11).

25.12 *Thm.* A is a formula & x is a variable $\rightarrow$ $D'_{25\,12}(A, x)$: $0 \tilde{\vdash} \tilde{\forall} x A \tilde{\rightarrow} A$.

Proof. Define $D' = \langle \tilde{\neg} A \tilde{\rightarrow} \tilde{\exists} x \tilde{\neg} A \rangle * \langle \tilde{\forall} x A \tilde{\rightarrow} A \rangle$.

25.13 *Thm.* (*substitution theorem*) a' is simultaneously substitutable for x' in $A \rightarrow C'_{25\,13}(a', x', A)$: $0 \tilde{\vdash} A_{x'}[a'] \tilde{\rightarrow} \text{Exist}\, x' * A$ &
$D'_{25\,13}(a', x', A)$: $0 \tilde{\vdash} \text{All}\, x' * A \tilde{\rightarrow} A_{x'}[a]$.

Proof. Define

$B' = \{\langle i, \text{Exist}\, x'[\text{Ln}\, x' - i + 2, \text{Ln}\, x'] * A \tilde{\rightarrow} \text{Exist}\, x'[\text{Ln}\, x' - i + 1] * A \rangle :$ $i \in \text{Dom}\, x'\}$,
$C' = B' * D'_{25\,6}(a', x', A \tilde{\rightarrow} \text{Exist}\, x' * A)$,
$B'_0 = \{\langle i, D'_{25\,12}(\text{All}\, x'[i, \text{Ln}\, x'] * A, x'(i)) \rangle : i \in \text{Dom}\, x'\}^*$,
$D' = B'_0 * D'_{25\,6}(a', x', \text{All}\, x' * A \tilde{\rightarrow} A)$.

Suppose *hyp* (25.13). Then $B \in \text{Ran}\, B' \rightarrow B$ is a substitutuion axiom, so B': $0 \tilde{\vdash} A \tilde{\rightarrow} \text{Exist}\, x' * A$. By (25.12) and (24.25), B'_0: $0 \tilde{\vdash} \text{All}\, x' * A \tilde{\rightarrow} A$. By the substitution rule (25.6) we have *con* (25.13), and thus (25.13).

25.14 *Thm.* (*distribution rule*) A and B are formulas & x is a variable $\rightarrow$
$C'_{25\,14}(A, B, x)$: $\langle A \tilde{\rightarrow} B \rangle \tilde{\vdash} \tilde{\exists} x A \tilde{\rightarrow} \tilde{\exists} x B$ &
$D'_{25.14}(A, B, x)$: $\langle A \tilde{\rightarrow} B \rangle \tilde{\vdash} \tilde{\forall} x A \tilde{\rightarrow} \tilde{\forall} x B$.

Proof. Define

$C' = \langle A \tilde{\rightarrow} B \rangle * \langle B \tilde{\rightarrow} \tilde{\exists} x B \rangle * \langle A \tilde{\rightarrow} \tilde{\exists} x B \rangle * \langle \tilde{\exists} x A \tilde{\rightarrow} \tilde{\exists} x B \rangle$,
$D' = D'_{25\,12}(A, x) * \langle \tilde{\forall} x A \tilde{\rightarrow} A \rangle * \langle A \tilde{\rightarrow} B \rangle * \langle \tilde{\forall} x A \tilde{\rightarrow} B \rangle * D'_{25.3}(\tilde{\forall} x A, B, x) *$ $\langle \tilde{\forall} x A \tilde{\rightarrow} \tilde{\forall} x B \rangle$.

25.15 *Def.* $\text{Free}\, A = \text{Enum}\, \{x \in \text{Occ}\, A : x \text{ is free in } A\}$.

25.16 *Def.* $\text{Closure}\, A = \text{All}\, \text{Free}\, A * A$.

25.17 *Thm.* (*closure theorem*) A is a formula $\to C'_{25\,17}(A)$: $\langle A\rangle \mathrel{\tilde{\vdash}}$ Closure A & $D'_{25\,17}(A)$: $\langle \text{Closure}\, A\rangle \mathrel{\tilde{\vdash}}$ A.

Proof. Define $x' = \text{Free}\, A$,

$$C' = \langle A\rangle * \{\langle i, D'_{25\,4}(\text{All}\, x'[i+1, \text{Ln}\, x'] * A, x'(i))\rangle : i \in \text{Dom}\, x\}^*,$$
$$D' = \langle \text{Closure}\, A\rangle * D'_{25\,13}(x', x', A) * \langle A\rangle. \ \square$$

Now we take up the deduction theorem; see [Sh,§3.3].

25.18 *Thm.* B is a tautological consequence of B' & A is a formula & $C' = \{\langle i, A \mathrel{\tilde{\to}} B'(i)\rangle : i \in \text{Dom}\, B'\} \to A \mathrel{\tilde{\to}} B$ is a tautological consequence of C'.

Proof. Suppose *hyp* (25.18), let $C = \text{Impl}(C' * \langle A \mathrel{\tilde{\to}} B\rangle)$, and suppose v is a truth valuation on C. Suppose 1: $v(B) = 1 \vee v(A) = 0$. Then $v(A \mathrel{\tilde{\to}} B) = 1$, so $v(C) = 1$ by (24.21). Thus (1) $\to v(C) = 1$, so suppose $v(B) = 0$ & $v(A) = 1$. By (24.21) there exists v_0 such that v_0 is a truth valuation on $\text{Impl}(B' * \langle B\rangle)$ & $\forall i(1 \le i \le \text{Ln}\, B' \to v_0(B'(i)) = v(B'(i)))$ & $v_0(B) = v(B)$. We have $v_0(\text{Impl}(B' * B)) = 1$, so by (24.21) there exists i such that $1 \le i \le \text{Ln}\, B'$ & $v_0(B'(i)) = 0$. Hence $v(B'(i)) = 0$, so $v(A \mathrel{\tilde{\to}} B'(i)) = 0$, and again by (24.21), $v(C) = 1$. Thus $v(C) = 1$, thus C is a tautology, and thus (25.18).

25.19 *Thm.* A and B are formulas & B': $A' \mathrel{\tilde{\vdash}} A \mathrel{\tilde{\to}} B \to$ $D'_{25\,19}(A, B, B')$: $A' * \langle A\rangle \mathrel{\tilde{\vdash}} B$.

Proof. Define $D' = \langle A\rangle * B' * \langle B\rangle$.

25.20 *Thm.* (*deduction theorem*) A is a closed formula & B': $A' * \langle A\rangle \mathrel{\tilde{\vdash}} B \to D'_{25\,20}(A, B')$: $A' \mathrel{\tilde{\vdash}} A \mathrel{\tilde{\to}} B$.

Proof. Define $C' = \{\langle i, A \mathrel{\tilde{\to}} B'(i)\rangle : i \in \text{Dom}\, B'\}$. We distinguish three cases:

1. $B'(i) \in \text{Ran}\, A' \vee B'(i)$ is a logical axiom,
2. $B'(i) = A \vee B'(i)$ is a tautological consequence of $B'[1, i-1]$,
3. $\exists j \exists x \exists C \exists D(x$ is a variable & C and D are formulas & $\neg(x$ is free in $D)$ & $1 \le j < i$ & $B'(i) = C \mathrel{\tilde{\to}} D$ & $B'(i) = \tilde{\exists} x C \mathrel{\tilde{\to}} D)$.

Define j, x, C, and D as in (3), and define

$$D'_1 = \langle B'(i)\rangle * \langle A \mathrel{\tilde{\to}} B'(i)\rangle,$$
$$D'_2 = \langle A \mathrel{\tilde{\to}} B'(i)\rangle,$$

$D'_3 = \langle A \tilde{\to} C \tilde{\to} D \rangle * \langle C \tilde{\to} A \tilde{\to} D \rangle * \langle \tilde{\exists} x C \tilde{\to} A \tilde{\to} D \rangle * \langle A \tilde{\to} \tilde{\exists} x C \tilde{\to} D \rangle$,
$D'_0 = D'_\mu$ in case (μ), for $1 \le \mu \le 3$,
$D' = \{\langle i, D'_0 \rangle : i \in \operatorname{Dom} B'\}^*$.

Suppose *hyp* (25.20), and suppose $1 \le i \le \operatorname{Ln} B'$. Observe that (3) $\to$ $\tilde{\exists} x C \tilde{\to} A \tilde{\to} D$ can be inferred from $C \tilde{\to} A \tilde{\to} D$ by $\exists$-introduction, since (3) $\to \neg(x$ is free in $A \tilde{\to} D)$. Thus $1 \le i \le \operatorname{Ln} B' \to D'_0$: $A' * C'[1, i-1] \,\tilde{\vdash}\, C'(i)$, and so *con* (25.20). Thus (25.20). □

How can one tell by looking at a finitary argument whether it admits a predicative arithmetization? A typical finitary argument, used to show that a certain kind of formula can be proved, introduces (perhaps implicitly) a ranking of formulas and reduces the proof for a given rank to the case of lower ranks, which are assumed to be already proved. To see whether such a reduction to previous cases can be constructed predicatively, we must examine the totality of previous cases and see if it is a set in the sense of our theory. There are three possible obstacles:

i. A splitting of cases may arise, leading to an exponential (or worse) growth in the total number of previous cases.

ii. There may be an exponential (or worse) growth in the size of the objects involved in the previous cases.

iii. The formula that specifies the previous cases may be unbounded.

Each application of the deduction theorem increases the length of a proof by a factor (which is ≤ 4), so if we try to convert a proof of a formula B from a sequence A' of n closed formulas into a proof of $\operatorname{Impl}(A' * \langle B \rangle)$ by induction on n, as in [Sh,§3.3], the length will grow exponentially. Here we have an example of the obstacle (ii). This induction is unbounded, and we must find another argument.

25.21 *Thm.* A' is a sequence of formulas & $1 \le i \le \operatorname{Ln} A' \to \operatorname{Conj} A' \tilde{\to} A'(i)$ is a tautology.

Proof. Suppose *hyp* (25.21) & v is a truth valuation on $\operatorname{Conj} A' \tilde{\to} A'(i)$ & $v(\operatorname{Conj} A' \tilde{\to} A'(i)) = 0$. Then $v(A'(i)) = 0$ and $v(\operatorname{Conj} A') = 1$. But this contradicts (24.22), and thus (25.21).

25.22 *Thm.* A' is a sequence of formulas & $A' \ne 0$ & A is a formula $\to$ $\operatorname{Impl}(A' * \langle A \rangle) \tilde{\to} (\operatorname{Conj} A' \tilde{\to} A)$ and $(\operatorname{Conj} A' \tilde{\to} A) \tilde{\to} \operatorname{Impl}(A' * \langle A \rangle)$ are tautologies.

Proof. Suppose *hyp* (25.22), let $C = \operatorname{Impl}(A' * \langle A \rangle)$, and let $D =$

$\operatorname{Conj} A' \stackrel{\sim}{\rightarrow} A$. Suppose v is a truth valuation on $C \stackrel{\sim}{\rightarrow} D$ & $v(C \stackrel{\sim}{\rightarrow} D) = 0$. Then $v(D) = 0$ and $v(C) = 1$. Therefore $v(A) = 0$ and $v(\operatorname{Conj} A') = 1$, so by (24.22), $\forall i(1 \leq i \leq \operatorname{Ln} A' \rightarrow v(A'(i)) = 1)$. By (24.21), $v(C) = 0$, a contradiction, and thus $C \stackrel{\sim}{\rightarrow} D$ is a tautology. Suppose v is a truth valuation on $D \stackrel{\sim}{\rightarrow} C$ & $v(D \stackrel{\sim}{\rightarrow} C) = 0$. Then $v(C) = 0$ and $v(D) = 1$. By (24.21), $v(A) = 0$ and $\forall i(1 \leq i \leq \operatorname{Ln} A' \rightarrow v(A'(i)) = 1)$. By (24.22), $v(D) = 0$, a contradiction, and thus $D \stackrel{\sim}{\rightarrow} C$ is a tautology. Thus (25.22).

25.23 *Def.* $\operatorname{Cases} A' = \{\langle i, \operatorname{Conj} A' \stackrel{\sim}{\rightarrow} A'(i)\rangle : i \in \operatorname{Dom} A'\}$.

25.24 *Thm.* A' is a sequence of formulas & B is a formula & B': $A'_0 \tilde{\vdash} \operatorname{Impl}(A' * \langle B \rangle) \rightarrow D'_{25\,24}(A', B, B')$: $A'_0 * A' \tilde{\vdash} B$.

Proof. Define $D' = A' * B' * \langle B \rangle$. Suppose *hyp* (25.24). Then *con* (25.24) by (25.23). Thus (25.24).

25.25 *Thm.* (*corollary to the deduction theorem*) A' is a sequence of closed formulas & B': $A'_0 * A' \tilde{\vdash} B \rightarrow D'_{25\,25}(A', B')$: $A'_0 \tilde{\vdash} \operatorname{Impl}(A' * \langle B \rangle)$ & $(A' \neq 0 \rightarrow C'_{25\,25}(A', B')$: $A'_0 \tilde{\vdash} \operatorname{Conj} A' \stackrel{\sim}{\rightarrow} B)$.

Proof. We distinguish two cases: 1. $A' = 0$ and 2. $A' \neq 0$. Define

$C' = D'_{25\,20}(\operatorname{Conj} A', \operatorname{Cases} A' * B')$,
$D'_1 = B'$,
$D'_2 = C' * \langle (\operatorname{Conj} A' \stackrel{\sim}{\rightarrow} B) \stackrel{\sim}{\rightarrow} \operatorname{Impl}(A' * \langle B \rangle)\rangle * \langle \operatorname{Impl}(A' * \langle B \rangle)\rangle$,
$D' = D'_\mu$ in case (μ), for $1 \leq \mu \leq 2$.

Suppose *hyp* (25.25). Then (1) $\rightarrow$ *con* (25.25), so suppose (2). By (25.21), $\operatorname{Cases} A' * B'$: $A'_0 * \langle \operatorname{Conj} A' \rangle \tilde{\vdash} B$. Then C': $A'_0 \tilde{\vdash} \operatorname{Conj} A' \stackrel{\sim}{\rightarrow} B$ by the deduction theorem (25.20). By (25.22), D'_2: $A'_0 \tilde{\vdash} \operatorname{Impl}(A' * \langle B \rangle)$, and thus (2) $\rightarrow$ *con* (25.25). Thus (25.25).

25.26 *Thm.* e' is a sequence of constants & x' is a sequence of variables & u is a designator $\rightarrow$ $(\operatorname{Ssub}(u, e', x')$ is a term $\leftrightarrow$ u is a term) & $(\operatorname{Ssub}(u, e', x')$ is a formula $\leftrightarrow$ u is a formula).

Proof. Suppose $\exists u \neg(25.26)$. By BLNP there exists a minimal such u. Let $u' = \operatorname{Arg} u$ and let

$$u'_1 = \{\langle i, \operatorname{Ssub}(u'(i), e', x')\rangle : i \in \operatorname{Dom} u'\}.$$

Then by the minimality assumption we have

$$\forall i(1 \le i \le \text{Ln}\, u_1' \rightarrow \textit{con}\,(25.26)_u[u_1'(i)]).$$

Suppose $\langle u(1)\rangle$ is a constant. Then Index $\langle u(1)\rangle = 0$, so $\text{Ln}\, u = 1$ and $u = \langle u(1)\rangle$. But then $\text{Ssub}(u, e', x')$ is a term, so *con* (25.26), a contradiction. Thus $\neg(\langle u(1)\rangle$ is a constant). Therefore $\text{Ssub}(u, e', x') = \langle u(1)\rangle * u_1'^{\,*}$ and so *con* (25.26), a contradiction. Thus (25.26).

25.27 *Thm.* e' is a sequence of constants & x' is a sequence of variables & A is a tautology $\rightarrow$ $\text{Ssub}(A, e', x')$ is a tautology.

Proof. Suppose *hyp* (25.27) and let $A_0 = \text{Ssub}(A, e', x')$. Suppose v_0 is a truth valuation on A_0 and let

$$v = \{\langle B, v_0(\text{Ssub}(B, e', x'))\rangle : B \in \text{Formulas}\, A\}.$$

Then v is a truth valuation on A, so $v(A) = 1$. Therefore $v_0(A_0) = 1$ and thus A_0 is a tautology. Thus (25.27).

25.28 *Thm.* (*theorem on constants*) e' is an injective sequence of constants & x' is an injective sequence of variables & $\text{Ln}\, x' = \text{Ln}\, e'$ & A' is a sequence of formulas & A is a formula & $\text{Ran}\, e' \cap \text{Occ}(A'^* * A) = 0$ & B': $A' \,\tilde{\vdash}\, A_{x'}[e'] \rightarrow D'_{25\,28}(e', x', A, B')$: $A' \,\tilde{\vdash}\, A$.

Proof. Define

$$\begin{aligned}
y' &= \{\langle i, X_{e'+x'+A+B'+i}\rangle : i \in \text{Dom}\, e'\},\\
C' &= \{\langle j, \text{Ssub}(B'(j), e', y')\rangle : j \in \text{Dom}\, B'\},\\
D' &= C' * D'_{25\,6}(x', y', A_{x'}[y']).
\end{aligned}$$

Suppose *hyp* (25.28). Observe that $\text{Ssub}(A_{x'}[e'], e', y') = A_{x'}[y']$. By (25.27), C': $A' \,\tilde{\vdash}\, A_{x'}[y']$. Since $(A_{x'}[y'])_{y'}[x'] = A$, we have *con* (25.28) by (25.6), and thus (25.28).

25.29 *Thm.* A and B are formulas & e' is an injective sequence of constants & $x' = \text{Free}\, A$ & $\text{Ln}\, x' = \text{Ln}\, e'$ & A' is a sequence of formulas & $\text{Ran}\, e' \cap \text{Occ}(A'^* * A * B) = 0$ & B': $A' * \langle A_{x'}[e']\rangle \,\tilde{\vdash}\, B_{x'}[e'] \rightarrow D'_{25\,29}(A, B, e', x', A', B')$: $A' \,\tilde{\vdash}\, A \rightleftarrows B$.

Proof. Define

$$\begin{aligned}
C' &= D'_{25\,20}(A_{x'}[e'], B', A'),\\
D' &= C' * D'_{25\,28}(e', x', A \rightleftarrows B, C').
\end{aligned}$$

Then (25.29) by the deduction theorem (25.20) and the theorem on constants (25.28). □

Now we take up the equivalence and equality theorems, following the discussion in [Sh,§3.4]. This is a dull topic; Shoenfield remarks that it can be roughly summarized by saying that equivalent formulas and equal terms may be substituted for one another.

We want to treat the notion of one formula being obtained from another by replacing some occurrences in it of formulas from a given sequence by the corresponding formula in another given sequence. We do this by means of another arithmetized relativized generalized inductive definition.

25.30 *Def.* $\text{Replacements}(A, A_1, B', B'_1) =$
$\text{Min}\, s(s \leq \text{Formulas}\, A \times \text{Formulas}\, A_1 \;\;\&$
$\forall B \forall B_1(\langle B, B_1\rangle \in \text{Formulas}\, A \times \text{Formulas}\, A_1 \;\;\&\;\; ((1) \vee \cdots \vee (5)) \rightarrow$
$\langle B, B_1\rangle \in s))$, *where*

1. $B = B_1$,
2. $\exists l(1 \leq l \leq \text{Ln}\, B' \;\;\&\;\; B = B'(l) \;\;\&\;\; B_1 = B'_1(l))$,
3. $\exists C \exists C_1(B = \tilde{\neg} C \;\&\; B_1 = \tilde{\neg} C_1 \;\&\; \langle C, C_1\rangle \in s)$,
4. $\exists C \exists D \exists C_1 \exists D_1(B = C \tilde{\vee} D \;\;\&\;\; B_1 = C_1 \tilde{\vee} D_1 \;\;\&\;\; \langle C, C_1\rangle \in s \;\;\&$ $\langle D, D_1\rangle \in s)$,
5. $\exists x \exists B \exists C(x \text{ is a variable} \;\;\&\;\; B = \tilde{\exists} x C \;\;\&\;\; B_1 = \tilde{\exists} x C_1 \;\;\&\;\; \langle C, C_1\rangle \in s)$.

25.31 *Def.* $\text{replace}(A, A_1, B', B'_1) \leftrightarrow A$ and B are formulas & B' and B'_1 are sequences of formulas & $\text{Ln}\, B' = \text{Ln}\, B'_1$ &
$\langle A, A_1\rangle \in \text{Replacements}(A, A_1, B', B'_1)$.

25.32 *Thm.* (*distribution rule for equivalence*) A and A_1 are formulas & x is a variable $\rightarrow C'_{25\,32}(A, A_1, x)$: $\langle A \tilde{\leftrightarrow} A_1\rangle \tilde{\vdash} \tilde{\exists} x A \tilde{\leftrightarrow} \tilde{\exists} x A_1$ &
$D'_{25\,32}(A, A_1, x)$: $\langle A \tilde{\leftrightarrow} A_1\rangle \tilde{\vdash} \tilde{\forall} x A \tilde{\leftrightarrow} \tilde{\forall} x A_1$.

Proof. Define

$C' = \langle A \tilde{\leftrightarrow} A_1\rangle * \langle A \tilde{\rightarrow} A_1\rangle * C'_{25\,14}(A, A_1, x) * \langle \tilde{\exists} x A \tilde{\rightarrow} \tilde{\exists} x A_1\rangle * \langle A_1 \tilde{\rightarrow} A\rangle *$
$C'_{25\,14}(A, A_1, x) * \langle \tilde{\exists} x A_1 \tilde{\rightarrow} \tilde{\exists} x A\rangle * \langle \tilde{\exists} x A \tilde{\leftrightarrow} \tilde{\exists} x A_1\rangle$,
$D' = \langle A \tilde{\leftrightarrow} A_1\rangle * \langle A \tilde{\rightarrow} A_1\rangle * D'_{25\,14}(A, A_1, x) * \langle \tilde{\forall} x A \tilde{\rightarrow} \tilde{\forall} x A_1\rangle * \langle A_1 \tilde{\rightarrow} A\rangle *$
$D'_{25\,14}(A, A_1, x) * \langle \tilde{\forall} x A_1 \tilde{\rightarrow} \tilde{\forall} x A\rangle * \langle \tilde{\forall} x A \tilde{\leftrightarrow} \tilde{\forall} x A_1\rangle$.

25.33 *Def.* $\text{Equiv}(B', B'_1) = \{\langle l, B'(l) \tilde{\leftrightarrow} B_1(l)\rangle : l \in \text{Dom}\, B'\}$.

25.34 *Thm.* (*equivalence theorem*) $\operatorname{replace}(A, A_1, B', B'_1) \to D'_{25.34}(A, A_1, B', B'_1)$: $\operatorname{Equiv}(B', B'_1) \mathrel{\tilde{\vdash}} A \mathrel{\tilde{\leftrightarrow}} A_1$.

Proof. Define

$s_0 = \operatorname{Ran}\{\langle z, \operatorname{Proj}_1 z \mathrel{\tilde{\leftrightarrow}} \operatorname{Proj}_2 z\rangle : z \in \operatorname{Replacements}(A, A_1, B', B'_1)\}$,
$r = \{\langle A_0, \operatorname{Ln} A_0\rangle : A_0 \in s_0\}$,
$A' = \operatorname{Enumer}(s_0, r)$.

We distinguish two cases:

1. $\exists x \exists C \exists C_1(x$ is a variable & $\langle C, C_1\rangle \in \operatorname{Replacements}(A, A_1, B', B'_1)$ & $A'(i) = \tilde{\exists} x C \mathrel{\tilde{\leftrightarrow}} \tilde{\exists} x C_1)$,
2. $\neg(1)$.

Define x, C, and C_1 as in (1), and define

$C'_1 = C'_{25.32}(C, C_1, x)$,
$C'_2 = \langle A'(i)\rangle$,
$C' = C'_\mu$ in case (μ), for $1 \le \mu \le 2$,
$D' = \{\langle i, C'\rangle : i \in \operatorname{Dom} A'\}^*$.

Suppose *hyp* (25.34) and suppose $1 \le i \le \operatorname{Ln} A'$. We have $(1) \to C'_1$: $\operatorname{Equiv}(B', B'_1) * A'[1, i-1] \mathrel{\tilde{\vdash}} A'(i)$ by (25.32). We have $(2) \to A'(i) \in \operatorname{Ran}\operatorname{Equiv}(B', B'_1) \vee A'(i)$ is a tautological consequence of $A'[1, i-1]$, so $(2) \to C'_2$: $\operatorname{Equiv}(B', B'_1) * A[1, i-1] \mathrel{\tilde{\vdash}} A'(i)$. Thus *con* (25.34), and thus (25.34). □

Using semibounded replacement, we can prove $\operatorname{replace}(A, A_1, B', B'_1)$ & $\forall l(1 \le l \le \operatorname{Ln} B' \to B'(l) \mathrel{\tilde{\leftrightarrow}} B'_1(l)$ is a theorem of $t) \to A \mathrel{\tilde{\leftrightarrow}} A_1$ is a theorem of t.

25.35 *Thm.* B is a formula & x and y are variables & $(x = y \vee \neg(y$ is free in $B)) \to D'_{25.35}(B, x, y)$: $0 \mathrel{\tilde{\vdash}} \tilde{\exists} x B \mathrel{\tilde{\leftrightarrow}} \tilde{\exists} y B_x[y]$.

Proof. Define

$D' = \langle B_x[y] \mathrel{\tilde{\rightarrow}} \tilde{\exists} x B\rangle * \langle \tilde{\exists} y B_x[y] \mathrel{\tilde{\rightarrow}} \tilde{\exists} x B\rangle * \langle B \mathrel{\tilde{\rightarrow}} \tilde{\exists} y B_x[y]\rangle * \langle \tilde{\exists} x B \mathrel{\tilde{\rightarrow}} \tilde{\exists} y B_x[y]\rangle * \langle \tilde{\exists} x B \mathrel{\tilde{\leftrightarrow}} \tilde{\exists} y B_x[y]\rangle$.

Suppose *hyp* (25.35). Then y is substitutable for x in B. Observe that $B \mathrel{\tilde{\rightarrow}} \tilde{\exists} y B_x[y]$ is a substitution axiom, since $B = (B_x[y])_y[x]$. Thus (25.35).

25.36 *Def.* $\operatorname{replace}_1(A, A_1, D, D_1) \leftrightarrow A$, A_1, D, and D_1 are formulas & $\exists i \exists j(1 \le i \le j \le \operatorname{Ln} A$ & $D = A[i, j]$ & $A_1 = A[1, i-1] * D_1 * A[j+1, \operatorname{Ln} A])$.

25.37 *Thm.* $C \in \text{Formulas}\, A \;\;\&\;\; C_1 \in \text{Formulas}\, A_1 \rightarrow$ $\text{Replacements}(C, C_1, B', B'_1) \subseteq \text{Replacements}(A, A_1, B', B'_1)$.

Proof. Suppose *hyp* (25.37), let $s_1 = \text{Replacements}(C, C_1, B', B'_1)$, and let $s = \text{Replacements}(A, A_1, B', B'_1)$, so $scope_{\text{Min}}\,(25.30)_{A,A_1,s}[C, C_1, s \cap s_1]$ and $s_1 \le s \cap s_1 \le s_1$. Thus (25.37).

25.38 *Thm.* $\text{replace}_1(A, A_1, D, D_1) \rightarrow \text{replace}(A, A_1, \langle D\rangle, \langle D_1\rangle)$.

Proof. Suppose $\exists A \exists A_1 \neg(25.38)$. By BLNP there exist minimal such A and A_1. Clearly $D \ne A$. By the occurrence theorem (23.44), there exists k such that $1 \le k \le \text{Index}\,\langle A(1)\rangle \;\;\&\;\; D$ occurs in $(\text{Arg}\, A)(k)$. By the minimality assumption,

$$\text{replace}((\text{Arg}\, A)(k), (\text{Arg}\, A_1)(k), \langle D\rangle, \langle D_1\rangle).$$

By (25.37) we have (25.38), a contradiction, and thus (25.38).

25.39 *Def.* A_1 is an immediate variant of $A \leftrightarrow \exists x \exists y \exists B(x$ and y are variables & B is a formula & x occurs in A & y occurs in A_1 & B occurs in A & $(x = y \vee \neg(y$ is free in $B))$ & $\text{replace}_1(A, A_1, \tilde{\exists} x B, \tilde{\exists} y B_x\lceil y\rceil))$.

25.40 *Thm.* A_1 is an immediate variant of $A \rightarrow D'_{25\,40}(A_1, A)\colon 0 \mathrel{\tilde{\vdash}} A \mathrel{\tilde{\leftrightarrows}} A_1$.

Proof. Define x, y, and B as in (25.39) and define

$D' = D'_{25\,35}(B, x, y) * D'_{25.34}(A, A_1, \langle\tilde{\exists} x B\rangle, \langle\tilde{\exists} y B_x\lceil y\rceil\rangle)$.

25.41 *Def.* A_1 is a variant of $A \leftrightarrow \exists A'(A'$ is a sequence of formulas & $1 \le \text{Ln}\, A' \le \text{Ln}\, A$ & $A'(1) = A$ & $A'(\text{Ln}\, A') = A_1$ & $\forall i(1 \le i \le \text{Ln}\, A'$ $\rightarrow \text{Ran}\, A'(i) \subseteq \text{Ran}\, A \cup \text{Ran}\, A_1$ & $\text{Ln}\, A'(i) = \text{Ln}\, A)$ & $\forall i(1 \le i < \text{Ln}\, A'$ $\rightarrow A'(i+1)$ is an immediate variant of $A'(i)))$.

We have *rhs* (25.41): $\text{Ln}\, A' \le \text{Ln}\, A$, $\text{Ln}\,\text{Sup}\, A' \le \text{Ln}\, A$, $\text{Sup}\,\text{Sup}\, A' \le \text{Max}(\text{Sup}\, A, \text{Sup}\, A_1)$.

25.42 *Thm.* A' is a sequence of formulas & $A' \ne 0 \rightarrow A'(1) \mathrel{\tilde{\leftrightarrows}} A'(\text{Ln}\, A')$ is a tautological consequence of $\text{Equiv}(A'[1, \text{Ln}\, A' - 1], A'[2, \text{Ln}\, A'])$.

Proof. Suppose $\exists A' \neg(25.42)$. By BLNP there exists a minimal such A'. Clearly $\text{Ln}\, A' \ne 1$. Let $A'_0 = A'[1, \text{Ln}\, A' - 1]$. By the minimality assumption we have $(25.42)_{A'}[A'_0]$; that is, $A'(1) \mathrel{\tilde{\leftrightarrows}} A'(\text{Ln}\, A' - 1)$ is a tautological consequence of $\text{Equiv}(A'[1, \text{Ln}\, A' - 2], A'[2, \text{Ln}\, A' - 1])$. But $A'(1) \mathrel{\tilde{\leftrightarrows}} A'(\text{Ln}\, A')$ is a tautological consequence of

$$\langle A'(1) \mathrel{\tilde{\leftrightarrow}} A'(\mathrm{Ln}\, A' - 1)\rangle * \langle A'(\mathrm{Ln}\, A' - 1) \mathrel{\tilde{\leftrightarrow}} A'(\mathrm{Ln}\, A')\rangle,$$

so *con* (25.42), a contradiction, and thus (25.42).

25.43 *Thm.* (*variant theorem*) A_1 is a variant of $A \rightarrow$ $D'_{25\,43}(A_1, A)$: $0 \mathrel{\tilde{\vdash}} A \mathrel{\tilde{\leftrightarrow}} A_1$.

Proof. Define A' as in (25.41) and define

$$D' = \{\langle i, D'_{25.40}(A'(i+1), A'(i))\rangle : \ i \in \mathrm{Dom}\, A'[1, \mathrm{Ln}\, A' - 1]\}^* * \langle A \mathrel{\tilde{\leftrightarrow}} A_1\rangle.$$

25.44 *Thm.* A is a formula & u is an expression $\rightarrow B_{25\,44}(A, u)$ is a variant of A & $\forall x(x \in \mathrm{Occ}\, u \rightarrow \neg(x$ is bound in $B_{25\,44}(A, u))$) & $\forall v(v \in \mathrm{Occ}\, u \rightarrow B_{25\,44}(A, u) = B_{25\,44}(A, v * u))$.

Proof. Define

$i' = \mathrm{Enum}\,\{i \in \mathrm{Dom}\, A : A'(i) \in \mathrm{Ran}\, u \ \ \& \ \ \langle A(i-1)\rangle = \tilde{\exists}\}$,
$j' = \{\langle k, j\rangle : k \in \mathrm{Dom}\, i' \ \ \& \ \ A[i'(k) + 1, j]$ is a formula$\}$,
$C' = \{\langle k, A[i'(k) + 1, j'(k)]\rangle : k \in \mathrm{Dom}\, i'\}$,
$x' = \{\langle k, \langle A(i'(k))\rangle\rangle : k \in \mathrm{Dom}\, i'\}$,
$s = \{x \in \mathrm{Occ}\, u : x$ is a variable$\}$,
$y' = \{\langle k, X_{A+s+k}\rangle : k \in \mathrm{Dom}\, i'\}$,
1. $A' = \mathrm{Min}\, A'(\mathrm{Ln}\, A' = \mathrm{Ln}\, i' + 1 \ \ \& \ \ A'(1) = A \ \ \&$
$\forall k(1 \leq k \leq \mathrm{Ln}\, i' \rightarrow A'(k+1) =$
$A'(k)[1, i'(k) - 2] * \tilde{\exists} * y'(k) * C'(k)_{x'(k)}\lceil y'(k)\rceil * A'(k)[j'(k) + 1, \mathrm{Ln}\, A]))$,
$B' = A'(\mathrm{Ln}\, A')$.

Suppose *hyp* (25.44). Then $scope_{\mathrm{Min}}$ (1): $\mathrm{Ln}\, A' \leq \mathrm{Ln}\, i' + 1$, $\mathrm{Ln}\,\mathrm{Sup}\, A' \leq \mathrm{Ln}\, A$, $\mathrm{Sup}\,\mathrm{Sup}\, A' \leq \mathrm{Max}(\mathrm{Sup}\, A, \mathrm{Sup}\, y')$. Suppose $1 \leq k \leq \mathrm{Ln}\, i'$. Then it follows that $A'(k+1)$ is an immediate variant of $A'(k)$. Thus B is a variant of A, and thus (25.44).

25.45 *Thm.* (*symmetry theorem*) a and b are terms $\rightarrow$ $D'_{25\,45}(a, b)$: $0 \mathrel{\tilde{\vdash}} a \mathrel{\tilde{=}} b \mathrel{\tilde{\leftrightarrow}} b \mathrel{\tilde{=}} a$.

Proof. Define

$$D' = \langle X_0 \mathrel{\tilde{=}} X_1 \mathrel{\tilde{\rightarrow}} X_0 \mathrel{\tilde{=}} X_0 \mathrel{\tilde{\rightarrow}} X_0 \mathrel{\tilde{=}} X_0 \mathrel{\tilde{\rightarrow}} X_1 \mathrel{\tilde{=}} X_0\rangle * \langle X_0 \mathrel{\tilde{=}} X_0\rangle * \langle X_0 \mathrel{\tilde{=}} X_1 \mathrel{\tilde{\rightarrow}} X_1 \mathrel{\tilde{=}} X_0\rangle * D'_{25\,8}(a \mathrel{\tilde{=}} b \mathrel{\tilde{\rightarrow}} b \mathrel{\tilde{=}} a, X_0 \mathrel{\tilde{=}} X_1 \mathrel{\tilde{\rightarrow}} X_1 \mathrel{\tilde{=}} X_0) * \langle a \mathrel{\tilde{=}} b \mathrel{\tilde{\rightarrow}} b \mathrel{\tilde{=}} a\rangle * D'_{25\,8}(b \mathrel{\tilde{=}} a \mathrel{\tilde{\rightarrow}} a \mathrel{\tilde{=}} b, X_0 \mathrel{\tilde{=}} X_1 \mathrel{\tilde{\rightarrow}} X_1 \mathrel{\tilde{=}} X_0) * \langle b \mathrel{\tilde{=}} a \mathrel{\tilde{\rightarrow}} a \mathrel{\tilde{=}} b\rangle * \langle a \mathrel{\tilde{=}} b \mathrel{\tilde{\leftrightarrow}} b \mathrel{\tilde{=}} a\rangle.$$

25.46 *Thm.* a is a term $\rightarrow D'_{25\,46}(a)$: $0 \mathrel{\tilde{\vdash}} a \mathrel{\tilde{=}} a$.

Proof. Define $D' = \langle X_0 \mathrel{\tilde{=}} X_0\rangle * D'_{25\,8}(a \mathrel{\tilde{=}} a, X_0 \mathrel{\tilde{=}} X_0) * \langle a \mathrel{\tilde{=}} a\rangle$.

25.47 *Thm.* f is a function synbol & c' and c'_1 are sequences of terms & $\text{Index}\, f = \text{Ln}\, c' = \text{Ln}\, c'_1 \to D'_{25.7}(f, c', c'_1)$: $\text{Equals}(c', c'_1) \tilde{\vdash} f{*}c'^{*} \cong f{*}c'^{*}_1$.

Proof. Define

$x' = \{\langle \imath, X_\imath \rangle i \in \text{Dom}\, c'\}$,
$x'_1 = \{\langle i, X_{\text{Ln}\, c' + \imath} : i \in \text{Dom}\, c'_1\}$,
$D_0 = \text{Impl}(\text{Equals}(x', x'_1) * \langle f{*}x'^{*} \cong f{*}x'^{*}_1 \rangle)$,
$D' = \text{Equals}(c', c'_1) * \langle D_0 \rangle * D'_{25\,6}(c' * c'_1, x' * x'_1, D_0) * \langle f{*}c'^{*} \cong f{*}c'^{*}_1 \rangle$.

25.48 *Thm.* p is a predicate symbol & c' and c'_1 are sequences of terms & $\text{Index}\, p = \text{Ln}\, c' = \text{Ln}\, c'_1 \to D'_{25\,48}(p, c', c'_1)$: $\text{Equals}(c', c'_1) \tilde{\vdash} p{*}c'^{*} \rightleftarrows p{*}c'^{*}_1$.

Proof. Define

$x' = \{\langle i, X_\imath \rangle : i \in \text{Dom}\, c'\}$,
$x'_1 = \{\langle i, X_{\text{Ln}\, c' + \imath} \rangle : i \in \text{Dom}\, c'_1\}$,
$D_0 = \text{Impl}(\text{Equals}(x', x'_1) * \langle p{*}x'^{*} \rightleftarrows p{*}x'^{*}_1 \rangle)$,
$D'_1 = \text{Equals}(c', c'_1) * \langle D_0 \rangle * D'_{25\,6}(c' * c'_1, x' * x'_1, D_0) * \langle p{*}c'^{*} \rightleftarrows p{*}c'^{*}_1 \rangle$,
$D'_2 = \{\langle i, D'_{25.45}(c'(i), c'_1(i)) \rangle : i \in \text{Dom}\, c'\}^{*}$,
$D' = D'_1 * D'_2 * (D'_1)_{c'c'_1}[c'_1 c'] * \langle p{*}c'^{*} \rightleftarrows p{*}c'^{*}_1 \rangle$.

25.49 *Def.* $\text{Substitutions}(u, u_1, a', a'_1) = \text{Min}\, s(s \leq \text{Occ}\, u \times \text{Occ}\, u_1$ & $\forall v \forall v_1(\langle v, v_1 \rangle \in \text{Occ}\, u \times \text{Occ}\, u_1$ & v and v_1 are designators & $((1) \vee (2) \vee (3)) \to \langle v, v_1 \rangle \in s))$, *where*

1. $v = v_1$,
2. $\exists l(1 \leq l \leq \text{Ln}\, a'$ & $v = a'(l)$ & $v_1 = a'_1(l))$,
3. $v(1) = v_1(1)$ & $(\langle v(1) \rangle = \overline{\exists} \to v(2) = v_1(2))$ &

$\forall i(1 \leq i \leq \text{Ln}\,\text{Index}\, \langle v(1) \rangle \to \langle (\text{Arg}\, v)(i), (\text{Arg}\, v_1)(i) \rangle \in s)$.

25.50 *Def.* $\text{substitute}(u, u_1, a', a'_1) \leftrightarrow u$ and u_1 are designators & a' and a'_1 are sequences of terms & $\text{Ln}\, a' = \text{Ln}\, a'_1$ &
$\langle u, u_1 \rangle \in \text{Substitutions}(u, u_1, a', a'_1)$.

25.51 *Thm.* (*equality theorem for terms*) b and b_1 are terms & $\text{substitute}(b, b_1, a', a'_1) \to D'_{25\,51}(b, b_1, a', a'_1)$: $\text{Equals}(a', a'_1) \tilde{\vdash} b \cong b_1$.

Proof. Define

$s_0 = \text{Ran}\, \{\langle z, \text{Proj}_1\, z \cong \text{Proj}_2\, z \rangle : z \in \text{Substitutions}(b, b_1, a', a'_1)\}$,
$r = \{\langle A_0, \text{Ln}\, A_0 \rangle : A_0 \in s_0\}$,
$A' = \text{Enumer}(s_0, r)$,

$$c' = \{\langle i, c\rangle : i \in \mathrm{Dom}\, A' \ \ \& \ \ \exists c_1\, A'(i) = c \,\hat{=}\, c_1\},$$
$$c_1' = \{\langle i, c_1\rangle : i \in \mathrm{Dom}\, A' \ \ \& \ \ \exists c\, A'(i) = c \,\hat{=}\, c_1\}.$$

We distinguish three cases:

1. $c'(i) = c_1'(i)$,
2. $c'(i) \hat{=} c_1(i) \in \mathrm{Ran\,Equals}(a', a_1')$,
3. $\neg(1)\ \&\ \neg(2)$.

Define

$$C_1' = D'_{2546}(c'(i)),$$
$$C_2' = \langle A'(i)\rangle,$$
$$C_3' = D'_{25\,47}(c'(i)(1), \mathrm{Arg}\,(c'(i)), \mathrm{Arg}\,(c_1'(i))),$$
$$C' = C_\mu' \text{ in case } (\mu), \text{ for } 1 \le \mu \le 3,$$
$$D' = \{\langle i, C'\rangle : i \in \mathrm{Dom}\, A'\}^*.$$

25.52 *Thm.* a' and a_1' are sequences of terms & $\mathrm{Ln}\, a' = \mathrm{Ln}\, a_1'$ & $\langle v, v_1\rangle \in$ $\mathrm{Substitutions}(u, u_1, a', a_1') \rightarrow$ (v and v_1 are terms $\vee$ v and v_1 are formulas) & substitute(v, v_1, a', a_1').

Proof. Suppose *hyp* (25.52), let $s = \mathrm{Substitutions}(u, u_1, a', a_1')$, and let

$$s_1 = \{z \in s : \mathit{con}\,(25.52)_{v,v_1}[\mathrm{Proj}_1\, z, \mathrm{Proj}_2\, z]\}.$$

Then we have $\mathit{scope}_{\mathrm{Min}}(25.49)_s[s_1]$, so $s_1 = s$. Thus (25.52).

25.53 *Def.* $u{\sim}v = A \leftrightarrow$ (u and v are terms & $A = u \hat{=} v$) $\vee$ (u and v are formulas & $A = u \rightleftarrows v$), otherwise $A = 1$.

25.54 *Thm.* (*equality theorem for formulas*) A and A_1 are formulas & substitute$(A, A_1, a', a_1') \rightarrow D'_{25\,54}(A, A_1, a', a_1')$: $\mathrm{Equals}(a', a_1') \,\tilde{\vdash}\, A \rightleftarrows A_1$.

Proof. Define

$$s_0 = \mathrm{Ran}\,\{\langle z, \mathrm{Proj}_1\, z{\sim}\mathrm{Proj}_2\, z\rangle : z \in \ \mathrm{Substitutions}(A, A_1, a', a_1')\},$$
$$r = \{\langle A_0, \mathrm{Ln}\, A_0\rangle : A_0 \in s_0\},$$
$$A' = \mathrm{Enumer}(s_0, r),$$
$$v' = \{\langle i, v\rangle : i \in \mathrm{Dom}\, A' \ \&\ \exists v_1\, v{\sim}v_1\},$$
$$v_1' = \{\langle i, v_1\rangle : i \in \mathrm{Dom}\, A' \ \&\ \exists v\, v{\sim}v_1\}.$$

We distinguish four cases:

1. $v'(i)$ and $v_1'(i)$ are terms,

2. $(\neg(1)\ \&\ v'(i) = v'_1(i)) \vee \langle v'(i)(1)\rangle = \bar{\neg} \vee \langle v'(i)(1)\rangle = \bar{\vee}$,
3. $v'(i) \neq v'_1(i)\ \ \&\ \ \langle v'(i)(1)\rangle$ is a predicate symbol,
4. $v'(i) \neq v'_1(i)\ \ \&\ \ \langle v'(i)(1)\rangle = \bar{\exists}$.

Define

$$\begin{aligned} C'_1 &= D'_{25.51}(v'(i), v'_1(i), a', a'_1), \\ C'_2 &= \langle A'(i)\rangle, \\ C'_3 &= D'_{25.48}(\langle v'(i)(1), \operatorname{Arg}(v'(i)), \operatorname{Arg}(v'_1(i))\rangle), \\ C'_4 &= D'_{25.32}(v'(i)[3, \operatorname{Ln} v'(i)], v'_1(i)[3, \operatorname{Ln} v'_1(i)], \langle v'(i)(2)\rangle), \\ C' &= C'_\mu \text{ in case } (\mu), \text{ for } 1 \leq \mu \leq 4, \\ D' &= \{\langle i, C'\rangle : i \in \operatorname{Dom} A'\}^*. \end{aligned}$$

25.55 *Thm.* v occurs in u & v_1 occurs in $u_1 \rightarrow$ Substitutions(v, v_1, a', a'_1) $\subseteq$ Substitutions(u, u_1, a', a'_1).

Proof. Suppose *hyp* (25.55), let $s_1 =$ Substitutions(v, v_1, a', a'_1), and let $s =$ Substitutions(u, u_1, a', a'_1). Then $scope_{\mathrm{Min}}\,(25.49)_{u,u_1,s}[v, v_1, s \cap s_1]$, so it follows that $s_1 \leq s \cap s_1 \leq s_1$. Thus (25.55).

25.56 *Thm.* u is a designator & x' is an injective sequence of variables & a' and a'_1 are sequences of terms & $\operatorname{Ln} x' = \operatorname{Ln} a' = \operatorname{Ln} a'_1$ & (u is a formula $\rightarrow a'$ is simultaneously substitutable for x' in u & a'_1 is simultaneously substitutable for x' in u) $\rightarrow$ substitute$(u_{x'}\lceil a'\rceil, u_{x'}\lceil a'_1\rceil, a', a'_1)$.

Proof. Suppose $\exists u \exists x' \neg(25.56)$. By BLNP there exist minimal such u and x'. Let $v = u_{x'}\lceil a'\rceil$, let $v_1 = u_{x'}\lceil a'_1\rceil$, and let

$$s = \text{Substitutions}(v, v_1, a', a'_1).$$

Clearly $\neg$(u is a variable), so $v(1) = v_1(1) = u(1)$. Suppose $\neg(\langle u(1)\rangle = \bar{\exists}$ & $\langle u(2)\rangle \in \operatorname{Ran} x')$. Suppose $1 \leq i \leq \operatorname{Index}\langle u(1)\rangle$, let $w = ((\operatorname{Arg} u)(i))_{x'}\lceil a'\rceil$, and let $w_1 = ((\operatorname{Arg} u)(i))_{x'}\lceil a'_1\rceil$. By the minimality assumption we have substitute(w, w_1, a', a'_1). But w occurs in v and w_1 occurs in v_1, so by (25.55), $\langle w, w_1\rangle \in s$. Thus $\langle v, v_1\rangle \in s$, a contradiction, and thus $\langle u(1)\rangle = \bar{\exists}$ & $\langle u(2)\rangle \in \operatorname{Ran} x'$. There exists j such that $1 \leq j \leq \operatorname{Ln} x'$ & $x'(j) = \langle u(2)\rangle$. Let $y' = x'[1, j-1] * x'[j+1, \operatorname{Ln} x']$; by (19.38) we have $y' \leq x'$. Suppose $1 \leq i \leq \operatorname{Index}\langle u(1)\rangle$, let $w = ((\operatorname{Arg} u)(i))_{y'}\lceil a'\rceil$, and let $w_1 = ((\operatorname{Arg} u)(i))_{y'}\lceil a'_1\rceil$. Again we have substitute(w, w_1, a', a'_1), w occurs in v and w_1 occurs in v_1, so $\langle w, w_1\rangle \in s$. Thus $\langle v, v_1\rangle \in s$, a contradiction, and thus (25.56).

25.57 *Thm.* (*equality theorem*) *hyp* (25.56) $\to$
$D'_{25.57}(u, x', a', a'_1)$: $0 \tilde{\vdash} \mathrm{Impl}(\mathrm{Equals}(a', a'_1) * \langle u_{x'}\lceil a'\rceil \sim u_{x'}\lceil a'_1\rceil\rangle)$.

(See Corollaries 1 and 2 of [Sh,§3.4]. The proof given there seems to require a slight emendation, as given below, because the variables occurring in one of the terms may overlap with the variables being substituted.)
Proof. Define

$$y' = \mathrm{Enum}\{y \in \mathrm{Occ\,Conj\,Equals}(a', a'_1) : y \text{ is a variable}\},$$
$$z' = \mathrm{Enum}\{z \in \mathrm{Ran}\, y' : z \notin \mathrm{Ran}\, x'\},$$
$$e' = \{\langle j, F_{0,u+a'+a'_1+j}\rangle : j \in \mathrm{Dom}\, y'\},$$
$$d' = \{\langle j, d\rangle : j \in \mathrm{Dom}\, y' \ \& \ \forall i(y'(j) = x'(i) \to d = x'(i)) \ \& \ (y'(j) \notin \mathrm{Ran}\, x' \to d = e'(j))\},$$
$$c' = \{\langle l, a'(l)_{y'}\lceil e'\rceil\rangle : l \in \mathrm{Dom}\, a'\},$$
$$c'_1 = \{\langle l, a'_1(l)_{y'}\lceil e'\rceil\rangle : l \in \mathrm{Dom}\, a'_1\},$$
$$v = u_{z'}\lceil d'\rceil.$$

We distinguish two cases: 1. u is a term, and 2. u is a formula. Define

$$C'_1 = D'_{25\,51}(v_{x'}\lceil c'\rceil, v_{x'}\lceil c'_1\rceil, c', c'_1),$$
$$C'_2 = D'_{25\,54}(v_{x'}\lceil c'\rceil, v_{x'}\lceil c'_1\rceil, c', c'_1),$$
$$C' = C'_\mu \text{ in case } (\mu), \text{ for } 1 \le \mu \le 2,$$
$$D'_1 = C'_{25\,25}(\mathrm{Equals}(c', c'_1), C', 0),$$
$$A_0 = \mathrm{Conj\,Equals}(a', a'_1) \rightrightarrows u_{x'}\lceil a'\rceil \sim u_{x'}\lceil a_1\rceil,$$
$$A_1 = \mathrm{Impl}\,(\mathrm{Equals}(a', a'_1) * \langle u_{x'}\lceil a'\rceil \sim u_{x'}\lceil a_1\rceil\rangle),$$
$$D'_2 = D'_{25\,28}(e', y', A_0, D'_1).$$

Unfortunately, there is a trivial distinction of cases to make: 3. $a' = a'_1 = 0$, and 4. $\neg(3)$. Define

$$D'_3 = \langle u \sim v\rangle,$$
$$D'_4 = D'_2 * \langle A_0 \rightrightarrows A_1\rangle * \langle A_1\rangle,$$
$$D' = D'_\mu \text{ in case } (\mu), \text{ for } 3 \le \mu \le 4.$$

Suppose *hyp* (25.57). Clearly (3) $\to$ *con* (25.57), so suppose (4). By (25.56), (25.51), and (25.54), C': $\mathrm{Equals}(c', c'_1) \tilde{\vdash} u_{x'}\lceil c'\rceil \sim u_{x'}\lceil c'_1\rceil$. Let

$$C = \mathrm{Conj\,Equals}(c', c'_1) \rightrightarrows (u_{x'}\lceil c'\rceil \sim u_{x'}\lceil c'_1\rceil).$$

By the corollary (25.25) to the deduction theorem, D'_1: $0 \tilde{\vdash} C$. But $A_{0y'}\lceil e'\rceil = C$, so by the theorem on constants (25.28), D'_2: $0 \tilde{\vdash} A_0$. Then by (25.22), D'_4: $0 \tilde{\vdash} A_1$, but $D' = D'_4$. Thus *con* (25.57), and thus (25.57).

25.58 *Thm.* (*corollary to the equality theorem*) a is substitutable for x in A & $\neg(x$ occurs in $a) \rightarrow D'_{25.58}(a, x, A)$: $0 \tilde{\vdash} A_x\lceil a\rceil \tilde{\leftrightarrow} \tilde{\exists}x(x \tilde{=} a \tilde{\&} A)$.

Proof. Define

$D' = D'_{25.57}(A, \langle x\rangle, \langle x\rangle, \langle a\rangle) * \langle x\tilde{=}a\tilde{\rightarrow}(A\tilde{\leftrightarrow}A_x\lceil a\rceil)\rangle * \langle x\tilde{=}a\tilde{\&}A\tilde{\rightarrow}A_x\lceil a\rceil\rangle * \langle\tilde{\exists}x(x\tilde{=}a\tilde{\&}A)\tilde{\rightarrow}A_x\lceil a\rceil\rangle * \langle a\tilde{=}a\tilde{\&}A_x\lceil a\rceil\tilde{\rightarrow}\tilde{\exists}x(x\tilde{=}a\tilde{\&}A)\rangle * D'_{25.46}(a) * \langle a\tilde{=}a\rangle * \langle A_x\lceil a\rceil\tilde{\leftrightarrow}\tilde{\exists}x(x\tilde{=}a\tilde{\&}A)\rangle$. □

Finally, we discuss the prenex operations; see [Sh,§3.5].

25.59 *Thm.* x is a variable & B and C are formulas & $\neg(x$ is free in $C) \rightarrow ((1)$ & $\cdots$ & $(6))$, *where*

1. $D'_{1,25.59}(x, B)$: $0 \tilde{\vdash} \tilde{\neg}\tilde{\exists}xB\tilde{\leftrightarrow}\tilde{\forall}x\tilde{\neg}B$,
2. $D'_{2,25.59}(x, B)$: $0 \tilde{\vdash} \tilde{\neg}\tilde{\forall}xB\tilde{\leftrightarrow}\tilde{\exists}x\tilde{\neg}B$,
3. $D'_{3,25.59}(x, B, C)$: $0 \tilde{\vdash} \tilde{\exists}xB\tilde{\vee}C\tilde{\leftrightarrow}\tilde{\exists}x(B\tilde{\vee}C)$,
4. $D'_{4,25.59}(x, B, C)$: $0 \tilde{\vdash} \tilde{\forall}xB\tilde{\vee}C\tilde{\leftrightarrow}\tilde{\forall}x(B\tilde{\vee}C)$,
5. $D'_{5,25.59}(x, B, C)$: $0 \tilde{\vdash} C\tilde{\vee}\tilde{\exists}xB\tilde{\leftrightarrow}\tilde{\exists}x(C\tilde{\vee}B)$,
6. $D'_{6,25.59}(x, B, C)$: $0 \tilde{\vdash} C\tilde{\vee}\tilde{\forall}xB\tilde{\leftrightarrow}\tilde{\forall}x(C\tilde{\vee}B)$.

Proof. Define

$D'_1 = \langle B\tilde{\leftrightarrow}\tilde{\neg}\tilde{\neg}B\rangle * C'_{25.32}(B, \tilde{\neg}\tilde{\neg}B, x) * \langle\tilde{\exists}xB\tilde{\leftrightarrow}\tilde{\exists}x\tilde{\neg}\tilde{\neg}B\rangle * \langle\tilde{\neg}\tilde{\exists}xB\tilde{\leftrightarrow}\tilde{\forall}x\tilde{\neg}B\rangle$,

$D'_2 = \langle\tilde{\neg}\tilde{\forall}xB\tilde{\leftrightarrow}\tilde{\exists}x\tilde{\neg}B\rangle$,

$D'_3 = \langle B\tilde{\rightarrow}B\tilde{\vee}C\rangle * C'_{25.14}(B, B\tilde{\vee}C, x) * \langle\tilde{\exists}xB\tilde{\rightarrow}\tilde{\exists}x(B\tilde{\vee}C)\rangle * \langle B\tilde{\vee}C\tilde{\rightarrow}\tilde{\exists}x(B\tilde{\vee}C)\rangle * \langle C\tilde{\vee}\tilde{\exists}x(B\tilde{\vee}C)\rangle * \langle B\tilde{\rightarrow}\tilde{\exists}xB\rangle * \langle B\tilde{\vee}C\tilde{\rightarrow}\tilde{\exists}xB\tilde{\vee}C\rangle * \langle\tilde{\exists}x(B\tilde{\vee}C)\tilde{\rightarrow}\tilde{\exists}xB\tilde{\vee}C\rangle * \langle\tilde{\exists}xB\tilde{\vee}C\tilde{\leftrightarrow}\tilde{\exists}x(B\tilde{\vee}C)\rangle$,

$D'_4 = \langle B\tilde{\rightarrow}B\tilde{\vee}C\rangle * D'_{25.14}(B, B\tilde{\vee}C, x) * \langle\tilde{\forall}xB\tilde{\rightarrow}\tilde{\forall}x(B\tilde{\vee}C)\rangle * \langle C\tilde{\rightarrow}B\tilde{\vee}C\rangle * D'_{25.3}(C, B\tilde{\vee}C, x) * \langle C\tilde{\rightarrow}\tilde{\forall}x(B\tilde{\vee}C)\rangle * D'_{25.12}(B\tilde{\vee}C, x) * \langle\tilde{\forall}x(B\tilde{\vee}C)\tilde{\rightarrow}B\tilde{\vee}C\rangle * \langle\tilde{\forall}x(B\tilde{\vee}C)\tilde{\&}\tilde{\neg}C\tilde{\rightarrow}B\rangle * D'_{25.3}(\tilde{\forall}x(B\tilde{\vee}C)\tilde{\&}\tilde{\neg}C, B, x) * \langle\tilde{\forall}x(B\tilde{\vee}C)\tilde{\&}\tilde{\neg}C\tilde{\rightarrow}\tilde{\forall}xB\rangle * \langle\tilde{\forall}x(B\tilde{\vee}C)\tilde{\rightarrow}\tilde{\forall}xB\tilde{\vee}C\rangle * \langle\tilde{\forall}xB\tilde{\vee}C\tilde{\leftrightarrow}\tilde{\forall}x(B\tilde{\vee}C)\rangle$,

$D'_5 = \langle C\tilde{\vee}B\tilde{\leftrightarrow}B\tilde{\vee}C\rangle * C'_{25.32}(C\tilde{\vee}B, B\tilde{\vee}C, x) * \langle\tilde{\exists}x(C\tilde{\vee}B)\tilde{\leftrightarrow}\tilde{\exists}x(B\tilde{\vee}C)\rangle * D'_3 * \langle\tilde{\exists}xB\tilde{\vee}C\tilde{\leftrightarrow}\tilde{\exists}x(B\tilde{\vee}C)\rangle * \langle C\tilde{\vee}\tilde{\exists}xB\tilde{\leftrightarrow}\tilde{\exists}x(C\tilde{\vee}B)\rangle$,

$D'_6 = \langle C\tilde{\vee}B\tilde{\leftrightarrow}B\tilde{\vee}C\rangle * D'_{25.32}(C\tilde{\vee}B, B\tilde{\vee}C, x) * \langle\tilde{\forall}x(C\tilde{\vee}B)\tilde{\leftrightarrow}\tilde{\forall}x(B\tilde{\vee}C)\rangle * D'_4 * \langle\tilde{\forall}xB\tilde{\vee}C\tilde{\leftrightarrow}\tilde{\forall}x(B\tilde{\vee}C)\rangle * \langle C\tilde{\vee}\tilde{\forall}xB\tilde{\leftrightarrow}\tilde{\forall}x(C\tilde{\vee}B)\rangle$.

25.60 *Def.* A_1 is an immediate prenex transform of $A \leftrightarrow \exists x\exists y\exists B\exists C\exists D\exists D_1$ (x and y are variables & A, A_1, B, C, D, and D_1 are formulas & x occurs in A & y occurs in A_1 & B occurs in A & C occurs in A & $\neg(y$ is free in C) & y is substitutable for x in B & replace$_1(A, A_1, D, D_1)$ & $((1) \vee \cdots \vee (6)))$, *where*

1. $D = \tilde{\neg}\tilde{\exists}xB$ & $D_1 = \tilde{\forall}x\tilde{\neg}B$,
2. $D = \tilde{\neg}\tilde{\forall}xB$ & $D_1 = \tilde{\exists}x\tilde{\neg}B$,
3. $D = \tilde{\exists}xB\tilde{\vee}C$ & $D_1 = \tilde{\exists}y(B_x\lceil y\rceil\tilde{\vee}C)$,
4. $D = \tilde{\forall}xB\tilde{\vee}C$ & $D_1 = \tilde{\forall}y(B_x\lceil y\rceil\tilde{\vee}C)$,
5. $D = C\tilde{\vee}\tilde{\exists}xB$ & $D_1 = \tilde{\exists}y(C\tilde{\vee}B_x\lceil y\rceil)$,
6. $D = C\tilde{\vee}\tilde{\forall}xB$ & $D_1 = \tilde{\forall}y(C\tilde{\vee}B_x\lceil y\rceil)$.

25.61 *Thm.* A_1 is an immediate prenex transform of $A \rightarrow$
$D'_{25\,61}(A_1, A)$: $0 \tilde{\vdash} A\tilde{\leftrightarrows}A_1$.

Proof. Define x, y, B, C, D, and D_1 as in (25.60). We distinguish the six cases (1)–(6) of (25.60). Define

$D'_0 = D'_{25\,34}(A, A_1, \langle D\rangle, \langle D_1\rangle)$,
$D'_1 = D'_{1,25\,59}(x, B) * D'_0$,
$D'_2 = D'_{2,25\,59}(x, B) * D'_0$,
$C' = D'_{25\,40}(D_1, D)$,
$D'_3 = C' * D'_{3,25\,59}(y, B_x\lceil y\rceil, C) * D'_0$,
$D'_4 = C' * D'_{4,25\,59}(y, B_x\lceil y\rceil, C) * D'_0$,
$D'_5 = C' * D'_{5,25\,59}(y, B_x\lceil y\rceil, C) * D'_0$,
$D'_6 = C' * D'_{6,25\,59}(y, B_x\lceil y\rceil, C) * D'_0$,
$D' = D'_\mu$ in case (μ), for $1 \leq \mu \leq 6$. □

It is possible to express within our theory the notion of a formula being in prenex form. One can construct the prenex form of a formula via a bounded function symbol. Each of the transforms (1)–(6) pushes a logical connective to the right of a quantifier. It is necessary to define the notion of occurrence of a logical connective in such a way that the occurrences of negation in a universal quantifier are not counted, and to impose the restriction in (1) that $\langle B(1)\rangle \neq \tilde{\neg}$. Then we obtain the prenex form A_1 of A by a sequence A' of formulas, each being an immediate prenex transform of its predecessor, with $\operatorname{Ln} A' \leq (\operatorname{Ln} A)^2 + 1$. The formulas may grow slightly in length because a universal quantifier has length 4 while an existential quantifier has length 2, but we obtain the bounds

$$\operatorname{Ln} A' \leq \operatorname{Log}(2 \cdot (A \# A)),$$

$$\operatorname{Ln} \operatorname{Sup} A' \leq \operatorname{Log}(A \cdot (2 \cdot (A \# a))^2),$$

and

$$\operatorname{Sup}\operatorname{Sup} A' \leq \operatorname{Max}(\operatorname{Sup} A, \operatorname{Sup} A_1).$$

Finally, one can construct a bounded function symbol that is a proof from 0 of the equivalence of a formula with its prenex form. I do not think that it will be necessary for our purposes to carry out these constructions.

Chapter 26

Special constants

There is another important derived rule of inference. When we have proved $\exists$xA it is very useful to have a name for such an x. In our proofs we have been saying "there exists x such that A". More formally, for a closed instantiation $\exists$xA one can adjoin a new constant r, called a *special constant*, and the *special axiom* $\exists$xA $\rightarrow$ $\mathrm{A_x}\lceil\mathrm{r}\rceil$, and this device can be iterated. When $\exists$xA is not closed, it is necessary to treat its free variables as constants; this is a tacit use of the Theorem on Constants. In this chapter we will construct a predicative arithmetization of the method of special constants, following the presentation contained within [Sh,§4.2].

26.1 *Def.* B is a special axiom $\leftrightarrow$ $\exists x \exists A \exists r$($x$ is a variable & A is a formula & r is a constant & x occurs in A & $\neg$(r occurs in A) & $\tilde{\exists} x A$ is a closed formula & $B = \tilde{\exists} x A \mathrel{\tilde{\rightarrow}} A_x\lceil y\rceil$).

26.2 *Def.* Spvar $B = x \leftrightarrow \exists A \exists r\ \mathit{scope}_{\exists r}$ (26.1), otherwise $x = 1$.

26.3 *Def.* Spform $B = A \leftrightarrow \exists x \exists r\ \mathit{scope}_{\exists r}$ (26.1), otherwise $A = 1$.

26.4 *Def.* Spconst $B = r \leftrightarrow \exists x \exists A\ \mathit{scope}_{\exists r}$ (26.1), otherwise $r = 1$.

26.5 *Thm.* A and B are formulas & x and y are variables & $\neg$(y occurs in $A \tilde{\vee} B$) $\rightarrow$ $D'_{26.5}(A, B, x, y)$: $\langle(\tilde{\exists} x A \mathrel{\tilde{\rightarrow}} A_x\lceil y\rceil) \mathrel{\tilde{\rightarrow}} B\rangle \mathrel{\tilde{\vdash}} B$.

Proof. Define

$D' = \langle(\tilde{\exists} x A \mathrel{\tilde{\rightarrow}} A_x\lceil y\rceil) \mathrel{\tilde{\rightarrow}} B\rangle * \langle\tilde{\exists} y(\tilde{\exists} x A \mathrel{\tilde{\rightarrow}} A_x\lceil y\rceil) \mathrel{\tilde{\rightarrow}} B\rangle * D'_{25\ 35}(A, x, y) * \langle\tilde{\exists} x A \mathrel{\tilde{\rightarrow}} \tilde{\exists} y A_x\lceil y\rceil\rangle * D'_{5,25.59}(y, A_x\lceil y\rceil, \tilde{\neg}\tilde{\exists} x A) * \langle\tilde{\exists} y(\tilde{\exists} x A \mathrel{\tilde{\rightarrow}} A_x\lceil y\rceil)\rangle * \langle B\rangle$.

26.6 *Def.* $t[B'] = \langle \text{Lang}\, t \cup \text{Nls}\, B'^{*}, \text{Ax}\, t \cup \text{Ran}\, B' \rangle$.

26.7 *Thm.* t is a theory & B' is a sequence of formulas $\rightarrow t[B']$ is a theory.

Proof. From (26.6) and (24.14).

26.8 *Thm.* (*theorem on special constants*) t is a theory & B' is a sequence of special axioms & $\forall u(u$ is a nonlogical symbol & u occurs in B'^{*} & $u \notin \text{Lang}\, t \rightarrow \exists i(1 \leq i \leq \text{Ln}\, B'$ & $u = \text{Spconst}\, B'(i)))$ & $\forall i(1 \leq i \leq \text{Ln}\, B'$ $\rightarrow \text{Spconst}\, B'(i) \notin \text{Lang}\, t \cup \text{Occ}\, B'[i+1, \text{Ln}\, B']^{*})$ & A is a formula of $\text{Lang}\, t$ & D'_0 is a proof in $t[B']$ of $A \rightarrow D'_{26.8}(t, B', A, D'_0)$ is a proof in t of A.

Proof. Define

$$
\begin{aligned}
r' &= \{\langle i, \text{Spconst}\, B'(i)\rangle : i \in \text{Dom}\, B'\},\\
C'_0 &= \{\langle i, \text{Spform}\, B'(i)\rangle : i \in \text{Dom}\, B'\},\\
A'_0 &= \{\langle i, \text{Impl}(B'[i, \text{Ln}\, B'] * \langle A\rangle)\rangle : \ i \in \text{Dom}\, B' \cup \{\text{Ln}\, B' + 1\}\},\\
D'_1 &= D'_{25\ 25}(B', D'_0, \text{Enum}\, \text{Ax}\, t),\\
y' &= \{\langle i, X_{B'+A+i}\rangle : i \in \text{Dom}\, B'\},\\
C'_1 &= \{\langle i, \text{Ssub}(C'_0(i), r', y')\rangle : i \in \text{Dom}\, B'\},\\
A'_1 &= \{\langle i, \text{Ssub}(A'_0(i), r', y')\rangle : i \in \text{Dom}\, A'_0\},\\
D'_2 &= D'_{25\ 28}(r', y', A'_1(1), D'_1),\\
D'_3 &= D'_{26\ 5}(C'_1(i), A'_1(i+1), \text{Spvar}\, B'(i), y'(i)),\\
D' &= D'_2 * \{\langle i, D'_3\rangle : i \in \text{Dom}\, B'\}^{*}.
\end{aligned}
$$

Suppose *hyp* (26.8). Observe that

1. $\forall i \forall j(1 \leq i < j \leq \text{Ln}\, B' \rightarrow \neg(r'(i)$ occurs in $B'(j)))$.

Now B' is a sequence of closed formulas, so by the corollary (25.25) to the deduction theorem, D'_1 is a proof in t of $\text{Impl}(B' * \langle A\rangle)$. Observe that $\text{Impl}(B' * \langle A\rangle) = A'_0(1)$. By the theorem on constants (25.28), D'_2 is a proof in t of $A'_1(1)$. By (26.5), which is applicable by (1), $1 \leq i \leq \text{Ln}\, B'$ $\rightarrow D'_3$: $A'_1(i) \vdash A'_1(i+1)$, so D' is a proof in t of $A'_1(\text{Ln}\, B' + 1)$. But $A'_1(\text{Ln}\, B' + 1) = A$, and thus (26.8).

Chapter 27

Extensions by definition

In this chapter we will predicatively arithmetize the notions of extension by definition of a predicate symbol and extension by definition of a function symbol, and show how to construct bounded function symbols that translate proofs into the original theory. We will follow [Sh,§4.6] except for the proof that an extension by definition of a function symbol is conservative. We begin with some general properties of translation functions.

27.1 *Def.* g is a translation function on $s \leftrightarrow g$ is a function & s is a set of formulas & $\operatorname{Dom} g = s$ & $\forall B(B \in s \rightarrow g(B)$ is a formula & $\operatorname{Free} g(B) = \operatorname{Free} B)$ & (1) & (2) & (3), *where*

1. $\forall B(\tilde{\neg} B \in s \rightarrow g(\tilde{\neg} B) = \tilde{\neg} g(B))$,
2. $\forall B \forall C(B \tilde{\vee} C \in s \rightarrow g(B \tilde{\vee} C) = g(B) \tilde{\vee} g(C))$,
3. $\forall B \forall x(\tilde{\exists} x B \in s \rightarrow g(\tilde{\exists} x B) = \tilde{\exists} x g(B))$.

27.2 *Def.* $\operatorname{Atoms} A = \{B \in \operatorname{Formulas} A : B$ is an atomic formula$\}$.

27.3 *Thm.* g_1 and g_2 are translation functions on Formulas A & $g_{1\,|\,\operatorname{Atoms} A} = g_{2\,|\,\operatorname{Atoms} A} \rightarrow g_1 = g_2$.

Proof. Suppose *hyp* (27.3), and suppose $\exists B(g_1(B) \neq g_2(B))$. By BLNP there is a minimal such B. Clearly $B \in \operatorname{Formulas} A$. By (23.30), B is an atomic formula, which is impossible. Thus $g_1 = g_2$, and thus (27.3).

27.4 *Thm.* g_0 is a translation function on Atoms $A \rightarrow \exists g(g$ is a translation function on Formulas A & $g_0 \subseteq g$ & $\operatorname{Ln} \operatorname{Sup} g \leq \operatorname{Ln} A \cdot \operatorname{Sup}(\operatorname{Ln} \circ g_0) \leq$

$\mathrm{Log}(A \,\#\, g_0(\mathrm{Maxm}(\mathrm{Ln} \circ g_0)))$ & $\mathrm{Sup}\,\mathrm{Sup}\, g \le \mathrm{Max}(\mathrm{Sup}\, A, \mathrm{Sup}\,\mathrm{Sup}\, g_0))$.

Proof. We use the abbreviation α for

g is a translation function on s & $s \subseteq \mathrm{Formulas}\, A$ & $g_0 \subseteq g$ & $\forall B(B \in s \to \mathrm{Formulas}\, B \subseteq s)$.

Suppose *hyp* (27.4). We claim that

1. α & $B \in s \to \mathrm{Ln}\, g(B) \le \mathrm{Ln}\, B \cdot \mathrm{Sup}(\mathrm{Ln} \circ g_0)$ & $\mathrm{Sup}\, g(B) \le \mathrm{Max}(\mathrm{Sup}\, B, \mathrm{Sup}\,\mathrm{Sup}\, g_0)$.

Suppose $\exists B \neg(1)$. By BLNP there exists a minimal such B, but this is impossible by (23.30). Thus (1). Since

$\mathrm{Ln}\, B \cdot \mathrm{Sup}\,(\mathrm{Ln} \circ g_0) \le \mathrm{Log}(A \,\#\, g_0(\mathrm{Maxm}(\mathrm{Ln} \circ g_0)))$,

we have

$\exists g\, \alpha$: $\mathrm{Dom}\, g \le \mathrm{Formulas}\, A$, $\mathrm{Ln}\,\mathrm{Sup}\, g \le \mathrm{Log}(A \,\#\, g_0(\mathrm{Maxm}(\mathrm{Ln} \circ g_0)))$, $\mathrm{Sup}\,\mathrm{Sup}\, g \le \mathrm{Max}(\mathrm{Sup}\, A, \mathrm{Sup}\,\mathrm{Sup}\, g_0)$.

We have $\alpha_{s,g}[\mathrm{Atoms}\, A, g_0]$ and hence $\exists g\, \alpha_s[\mathrm{Atoms}\, A]$. Let

$s = \mathrm{Max}\, s(s \le \mathrm{Formulas}\, A \;\&\; \exists g\, \alpha)$.

Then $\exists g\, \alpha$ by MAX, so of course there exists g such that α. Suppose

2. $\exists B(B \in \mathrm{Formulas}\, A \;\&\; B \notin s)$.

By BLNP there exists a minimal such B. Let $s_1 = s \cup \{B\}$. Now $B \notin \mathrm{Atoms}\, A$, so by (23.30) there exist C, D, and x such that

$C \in s \;\&\; D \in s \;\&\; (B = \tilde{\neg} C \vee B = C \tilde{\vee} B \vee B = \tilde{\exists} x C)$.

Suppose $B = \tilde{\neg} C$ and let $g_1 = g \cup \{\langle \tilde{\neg} C, \tilde{\neg} g(C)\rangle\}$. Then $\alpha_{sg}[s_1 g_1]$, so $\exists g\, \alpha_s[s_1]$. By the maximality of s we have $s_1 \le s$, a contradiction. Thus $B \ne \tilde{\neg} C$. Suppose $B = C \tilde{\vee} D$ and let $g_1 = g \cup \{\langle C \tilde{\vee} D, g(C) \tilde{\vee} g(D)\rangle\}$. Then $s_1 \le s$, a contradiction, and thus $B = \tilde{\exists} x C$. Let $g_1 = g \cup \{\langle \tilde{\exists} x C, \tilde{\exists} x g(C)\rangle\}$. Then $s_1 \le s$, a contradiction, and thus $\neg(2)$. Therefore $s = \mathrm{Formulas}\, A$, and thus (27.4).

27.5 *Def.* $\mathrm{Trext}(g_0, A) = g \leftrightarrow g_0$ is a translation function on $\mathrm{Atoms}\, A$ & g is a translation function on $\mathrm{Formulas}\, A$ & $g_0 \subseteq g$, otherwise $g = 1$.

We have been using the abbreviation $g \circ f$. Now we introduce a binary function symbol denoted by $\circ$.

27.6 *Def.* $g \circ f = \{\langle x, z\rangle : x \in \operatorname{Dom} f \ \& \ f(x) \in \operatorname{Dom} g \ \& \ z = g(f(x))\}$.

27.7 *Thm.* g is a translation function on s & B' is a sequence of formulas & $\operatorname{Impl} B' \in s \rightarrow g(\operatorname{Impl} B') = \operatorname{Impl}(g \circ B')$.

Proof. Suppose $\exists B' \neg(27.7)$. By BLNP there exists a minimal such B'. Clearly $\operatorname{Ln} B' \geq 2$. Let $B'_1 = B'[2, \operatorname{Ln} B']$. Then $\operatorname{Impl} B' = B'(1) \tilde{\rightarrow} \operatorname{Impl} B'_1$, so *con* (27.7), a contradiction, and thus (27.7).

27.8 *Thm.* g is a translation function on s & $\operatorname{Impl}(B' * \langle B\rangle) \in s$ & B is a tautological consequence of $B' \rightarrow g(B)$ is a tautological consequence of $g \circ B'$.

(See the Remark in [Sh,§3.1].) *Proof.* Suppose *hyp* (27.8) and let $A = \operatorname{Impl}(B' * \langle B\rangle)$. By (27.7), $g(A) = \operatorname{Impl}(g \circ B' * \langle g(B)\rangle)$. Suppose v is a truth valuation on $g(A)$, and let

$$v_0 = \{\langle C, v(g(C))\rangle : C \in \operatorname{Formulas} A\}.$$

Then v_0 is a truth valuation on A, so $v_0(A) = 1$ and hence $v(A) = 1$. Thus $g(A)$ is a tautology, and thus (27.8).

27.9 *Thm.* g is a translation function on s & $C \in s$ & $D \in s$ & C can be inferred from D by $\exists$-introduction $\rightarrow g(C)$ can be inferred from $g(D)$ by $\exists$-introduction.

Proof. Suppose *hyp* (27.9). There exist x, A, and B such that $\neg$(x is free in B) & $D = A \tilde{\rightarrow} B$ & $C = \tilde{\exists} x A \tilde{\rightarrow} B$. Since $\operatorname{Free} g(B) = \operatorname{Free} B$, we have $\neg$(x is free in $g(B)$). Also, $g(D) = g(A) \tilde{\rightarrow} g(B)$ and $g(C) = \tilde{\exists} x g(A) \tilde{\rightarrow} g(B)$, so *con* (27.9). Thus (27.9).

27.10 *Thm.* A is a formula & g is a translation function on Formulas A $\rightarrow \operatorname{replace}(A, g(A), \operatorname{Enum} \operatorname{Atoms} A, g \circ \operatorname{Enum} \operatorname{Atoms} A)$.

Proof. Suppose *hyp* (27.10), and let $B' = \operatorname{Enum} \operatorname{Atoms} A$. Suppose 1: $\exists B(B \in \operatorname{Formulas} A \ \& \ \langle B, g(B)\rangle \notin \operatorname{Replacements}(A, g(A), B', g \circ B'))$. By BLNP there exists a minimal such B, which is impossible. Thus (1), so that *con* (27.10). Thus (27.10). □

Now we introduce the notion of an extension by definition of a predicate symbol.

27.11 *Def.* $\operatorname{extp}(t_1, t, p, x', D) \leftrightarrow t_1$ and t are theories & p is a predicate symbol & $p \notin \operatorname{Lang} t$ & $p \neq \equiv$ & x' is an injective sequence of variables

& Index p = Ln x' & D is a formula of Lang t & Ran Free $D \subseteq$ Ran x' & $t_1 = t[\langle p{*}x'^{*} \rightleftarrows D\rangle]$.

27.12 *Def.* $\text{Tratp}(p, x', D, u, A) = C \leftrightarrow ((\langle A(1)\rangle \neq p \rightarrow C = A)$ & $(\langle A(1)\rangle = p \rightarrow C = B_{25.44}(D, A * u)_{x'}[\text{Arg}\, A]).$

This is the arithmetization of A*, for A an atomic formula, in the discussion of predicate symbols in [Sh,§4.6]. Recall that $B_{25\,44}(D, A * u)$ is a variant of D, constructed to avoid colliding variables with $A * u$. It is convenient to allow the dependence on an auxiliary expression u; this will save a lot of fuss with variants.

27.13 *Def.* A is an atomic formula of $l \leftrightarrow A$ is an atomic formula & A is a formula of l.

27.14 *Thm.* $\text{extp}(t_1, t, p, x', D)$ & A is an atomic formula of Lang $t_1 \rightarrow$ $D'_{27.14}(p, x', D, A, u)$ is a proof in t_1 of $A \rightleftarrows \text{Tratp}(p, x', D, u, A)$.

(See (i) in the discussion of predicate symbols in [Sh,§4.6]. The restriction to atomic formulas will be removed shortly.) *Proof.* We distinguish two cases: 1. $\langle A(1)\rangle \neq p$ and 2. $\langle A(1)\rangle = p$. Define

$D'_1 = \langle A \rightleftarrows A\rangle$,
$D_0 = B_{25\,44}(D, A * u)$,
$D'_2 = \langle p{*}x'^{*} \rightleftarrows D\rangle * D'_{25\,43}(D_0, D) * \langle p{*}x'^{*} \rightleftarrows D_0\rangle *$
$D'_{25\,6}(\text{Arg}\, A, x', p{*}x'^{*} \rightleftarrows D_0) * \langle p{*}(\text{Arg}\, A)^{*} \rightleftarrows D_{0x'}[\text{Arg}\, A]\rangle$,
$D' = D'_\mu$ in case (μ), for $1 \leq \mu \leq 2$.

Suppose *hyp* (27.14). Clearly (1) $\rightarrow$ *con* (27.14), so suppose (2). Then $p{*}(\text{Arg}\, A)^{*} = A$ and $D_{0x'}[\text{Arg}\, A] = \text{Tratp}(p, x', D, u, A)$, so by the variant theorem (25.43) and the substitution rule (25.5) we have *con* (27.14). Thus *con* (27.14), and thus (27.14).

27.15 *Def.* $\text{Trfnp}(p, x', D, u, A) =$
$\text{Trext}(\{\langle B, \text{Tratp}(p, x', D, u, B)\rangle : B \in \text{Atoms}\, A\}, A)$.

27.16 *Def.* $\text{Trp}(p, x', D, u, A) = \text{Trfnp}(p, x', D, u, A)(A)$.

27.17 *Thm.* $\text{extp}(t_1, t, p, x', D)$ & A is a formula of Lang $t_1 \rightarrow$
$\text{Trp}(p, x', D, u, A)$ is a formula of Lang t & (A is a formula of Lang $t \rightarrow$
$\text{Trp}(p, x', D, u, A) = A$).

Proof. Suppose $\exists A \neg (27.17)$. By BLNP there exists a minimal such A, which is impossible. Thus (27.17).

27.18 *Thm.* *hyp* (27.17) $\rightarrow D'_{27\,18}(p, x', D, u, A)$ is a proof in t_1 of $A \leftrightarrows \mathrm{Trp}(p, x', D, u, A)$.

Proof. Define

$B'_1 = \mathrm{Enum\,Atoms}\, A$,
$D'_1 = \{\langle i, D'_{27\,14}(p, x', D, B'(i), u)\rangle : i \in \mathrm{Dom}\, B'\}^* *$
$D'_{25.34}(A, \mathrm{Trp}(p, x', D, u, A), B', B'_1)$.

Then (27.18) by (27.14) and the equivalence theorem (25.34), which is applicable by (27.10).

27.19 *Thm.* $g = \mathrm{Trfnp}(p, x', D, u, A)$ & x is a variable & a is a term & $B \in \mathrm{Formulas}\, A$ & $B_x[a] \in \mathrm{Formulas}\, A \rightarrow g(B_x[a]) = (g(B)_x[a])$.

Proof. Suppose $\exists B \neg(27.19)$. By BLNP there exists a minimal such B. Clearly $\neg(B$ is an atomic formula), $\langle B(1)\rangle \neq \tilde{\neg}$, and $\langle B(1)\rangle \neq \tilde{\vee}$. Therefore $\langle B(1)\rangle = \tilde{\exists}$. Let $y = \langle B(2)\rangle$ and let $C = B[3, \mathrm{Ln}\, B]$, so that $B = \tilde{\exists} yC$. Suppose $y = x$. Then $g(B_x[a]) = g(B) = (g(B)_x[a])$, a contradiction, and thus $y \neq x$. Then $g(B_x[a]) = g(\tilde{\exists} yC_x[a]) = \tilde{\exists} yg(C_x[a]) = \tilde{\exists} y(g(C))_x[a]$ by the minimality assumption, and $\tilde{\exists} y(g(C))_x[a] = (g(\tilde{\exists} yC))_x[a] = (g(B))_x[a]$. This is a contradiction, and thus (27.19).

27.20 *Thm.* $\mathrm{extp}(t_1, t, p, x', D)$ & D'_0 is a proof in t_1 of $A \rightarrow$ $D'_{27\,20}(p, x', D, D'_0)$ is a proof in t of $\mathrm{Trp}(p, x', D, D'^*_0, A)$.

(See (ii) in the discussion of predicate symbols in [Sh,§4.6].) *Proof.* Suppose *hyp* (27.20) and define

$g = \mathrm{Trfnp}(p, x', D, D'^*_0, \mathrm{Disj}\, D'_0)$,
$D_0 = B_{25\,44}(D, D'^*_0)$.

Observe that

$\forall i \forall B(1 \leq i \leq \mathrm{Ln}\, D'_0$ & $B \in \mathrm{Formulas}\, D'_0 \rightarrow$
$g(B) = \mathrm{Trp}(p, x', D, D'^*_0, B))$

by (27.3). Suppose $1 \leq i \leq \mathrm{Ln}\, D'_0$. We distinguish three cases:

1. $D'_0(i)$ is a substitution axiom $\vee$ $D'_0(i)$ is an identity axiom $\vee$ $(D'_0(i)$ is an equality axiom & $\neg(p$ occurs in $D'_0(i)))$ $\vee$ $D'_0(i) \in \mathrm{Ax}t$ $\vee$ $D'_0(i)$ is a tautological consequence of $D'_0[1, i-1]$ $\vee$ $\exists j(1 \leq j < i$ & $D'_0(i)$ can be inferred from $D'_0(j)$ by $\exists$-introduction),

2. $\neg(1)$ & $D'_0(i)$ is an equality axiom,

3. $\neg(1)$ & $D'_0(i) = p{*}x'^{*}\tilde{\leftrightarrow}D$.

Suppose (1) and define $C'_1 = \langle g(D'_0(i))\rangle$. We claim that

4. C'_1: $\operatorname{Enum}\operatorname{Ax} t * g \circ D'_0[1, i-1] \,\tilde{\vdash}\, g(D'_0(i))$.

This holds by (27.19) (for a substitution axiom), by (27.17) (for axioms of $\operatorname{Lang} t$), by (27.8) (for tautological consequence), and by (27.9) (for $\exists$-introduction). Thus $(1) \to (4)$.

Suppose (2), define y' and z' as in

$$\exists y' \exists z'\, D'_0(i) = \operatorname{Impl}(\operatorname{Equals}(y', z') * \langle p{*}y'^{*}\tilde{\rightarrow}p{*}z'^{*}\rangle),$$

and define $C'_2 = D'_{25\,57}(D_0, x', y', z')$. Thus $(2) \to (4)_{C'_1}[C'_2]$ by the corollary (25.57) to the equality theorem.

Suppose (3) and define $C'_3 = D'_{25\,43}(D_0, D) * \langle D_0\tilde{\leftrightarrow}D\rangle$. Observe that $D_0\tilde{\leftrightarrow}D = g(D'_0(i))$, and thus $(3) \to (4)_{C'_1}[C'_3]$ by the variant theorem (25.43).

Define $C' = C'_\mu$ in case (μ), for $1 \le \mu \le 3$. Thus $1 \le i \le \operatorname{Ln} D'_0 \to (4)_{C'_1}[C]$. Define $D' = \{\langle i, C'\rangle : i \in \operatorname{Dom} D'_0\}^{*}$. Then D' is a proof in t of $g(A)$, and thus (27.20). □

In arithmetizing the notion of an extension by definition of a function symbol, we need to include the proofs of the existence and uniqueness conditions in order to have a bounded predicate symbol. So as not to have to worry again about variants, we first treat the case that the right hand side of the defining axiom is an atomic formula; the general case is easily reduced to this by first adjoining a new predicate symbol.

27.21 *Def.* $\operatorname{extf}_0(t_1, t, f, x', y, y_1, p, C'_0, C'_1) \leftrightarrow t_1$ and t are theories & f is a function symbol & $f \notin \operatorname{Lang} t$ & $x' * \langle y\rangle * \langle y_1\rangle$ is an injective sequence of variables & $\operatorname{Index} f = \operatorname{Ln} x'$ & p is a predicate symbol & $\operatorname{Index} p = \operatorname{Index} f + 1$ & C'_0 is a proof in t of $\tilde{\exists} y p{*}x'^{*}{*}y$ & C'_1 is a proof in t of $p{*}x'^{*}{*}y\,\tilde{\&}\,p{*}x'^{*}{*}y_1\tilde{\rightarrow}y\tilde{=}y_1$ & $t_1 = t[y\tilde{=}f{*}x'^{*}\tilde{\leftrightarrow}p{*}x'^{*}{*}y]$.

Now we introduce some function symbols to be used in eliminating occurrences of f.

27.22 *Def.* $\operatorname{Newvar} u = \operatorname{Min} z(z \le X_u$ & z is a variable & $\neg(z$ occurs in $u))$.

27.23 *Def.* $\operatorname{Last}_1(f, u) = i \leftrightarrow \langle u(i)\rangle = f$ & $\forall j(i < j \le \operatorname{Ln} u \to \langle u(i)\rangle \ne f)$, otherwise $i = 0$.

27.24 *Def.* $\mathrm{Last}_2(f,u) = j \leftrightarrow u$ and $u[\mathrm{Last}_1(f,u), j]$ are designators, otherwise $j = 0$.

27.25 *Def.* $\mathrm{Arglast}(f,u) = \mathrm{Arg}(u[\mathrm{Last}_1(f,u), \mathrm{Last}_2(f,u)])$.

27.26 *Def.* $\mathrm{Elim}(f,u) = u[1, \mathrm{Last}_1(f,u) - 1] * \mathrm{Newvar}\, u *$ $u[\mathrm{Last}_2(f,u) + 1, \mathrm{Ln}\, u]$.

27.27 *Def.* $\mathrm{Deg}(f,u) = \mathrm{Card}\{i \in \mathrm{Dom}\, u : \langle u(i)\rangle = f\}$.

27.28 *Thm.* f is a function symbol & u is a designator & f occurs in u & $z = \mathrm{Newvar}\, u$ & $a' = \mathrm{Arglast}(f,u)$ & $v = \mathrm{Elim}(f,u) \rightarrow u = v_z\lceil f{*}a'^{*}\rceil$ & $\neg(f$ occurs in $a'^{*})$ & $\mathrm{Deg}(f,v) = \mathrm{Deg}(f,u) - 1$.

Proof. From (27.22)–(27.27).

27.29 *Thm.* f is a function symbol & u is a designator $\rightarrow \exists! v'(\mathrm{Ln}\, v' = \mathrm{Deg}(f,u)+1$ & $v'(1) = u$ & $\forall i(1 \le i < \mathrm{Ln}\, v' \rightarrow v'(i+1) = \mathrm{Elim}(f, v'(i)))$ & $\forall i(1 \le i \le \mathrm{Ln}\, v' \rightarrow \mathrm{Deg}(f, v'(i)) = \mathrm{Deg}(f,u)+i-1)$ & $\mathrm{Ln}\,\mathrm{Sup}\, v' \le \mathrm{Ln}\, u$ & $\mathrm{Sup}\,\mathrm{Sup}\, v' \le X_{u+\mathrm{Ln}\, u})$.

Proof. The uniqueness holds by (12.17). We use α as an abbreviation for $scope_{\exists! v'}$ (27.29) but with $\mathrm{Ln}\, v' = \mathrm{Deg}(f,u) + 1$ replaced by $\mathrm{Ln}\, v' \le \mathrm{Deg}(f,u) + 1$. Suppose *hyp* (27.29) and let $v' = \mathrm{Max}\, v'\, \alpha$. We have $\alpha_{v'}[\langle u\rangle]$, so by MAX we have α. Suppose f occurs in $v'(\mathrm{Ln}\, v')$, and let

$$v'_0 = v \cup \{\langle \mathrm{Ln}\, v' + 1, \mathrm{Elim}(f, v'(\mathrm{Ln}\, v'))\rangle\}.$$

Then $\alpha_{v'}[v'_0]$ by (27.28), so that $v'_0 \le v'$, which is an impossibility. Thus $\neg(f$ occurs in $v'(\mathrm{Ln}\, v'))$; that is, $\mathrm{Deg}(f, v'(\mathrm{Ln}\, v')) = 0$. Consequently, $\mathrm{Ln}\, v' = \mathrm{Deg}(f,u) + 1$, and thus (27.29).

27.30 *Def.* $\mathrm{Elimseq}(f,u) = v' \leftrightarrow$ *hyp* (27.29) & $scope_{\exists! v'}$ (27.29), otherwise $v' = 1$.

27.31 *Thm.* f is a function symbol & A is an atomic formula $\rightarrow \exists! g(g$ is a function & $\mathrm{Dom}\, g = \mathrm{Ran}\,\mathrm{Elimseq}(f,A)$ & $\forall B(B \in \mathrm{Dom}\, g \rightarrow (1)$ & $\cdots$ & $(4)))$, *where*

1. $\mathrm{Ln}\, g(B) \le \mathrm{Ln}\, B + (8 + \mathrm{Index}\, f) \cdot \mathrm{Deg}(f,B)$ & $\mathrm{Ln}\, g(b) \le \mathrm{Log}(B^9 \cdot (B \# B))$,
2. $\mathrm{Sup}\, g(B) \le X_{A+\mathrm{Ln}\, A} + p + \overline{\exists}$,
3. $\mathrm{Deg}(f,B) = 0 \rightarrow g(B) = B$,

4. $\forall z \forall C \forall a'(\mathrm{Deg}(f,B) \neq 0 \;\;\&\;\; z = \mathrm{Newvar}\, B \;\;\&\;\; C = \mathrm{Elim}(f,B) \;\;\&\;\; a' = \mathrm{Arglast}(f,B) \rightarrow g(B) = \tilde{\exists} z(p{*}a'^{*}{*}z \tilde{\&} g(C)))$.

Proof. Suppose *hyp* (27.31). Let $A = \mathrm{Reverse\,Elimseq}(f,A)$. Write α as an abbreviation for $scope_{\exists!g}$ (27.31) but with $\mathrm{Dom}\, g = \mathrm{Ran\,Elimseq}(f,A)$ replaced by $\exists j(1 \le j \le \mathrm{Ln}\, A' \;\;\&\;\; \mathrm{Dom}\, g = \mathrm{Ran}\, A'[1,j])$, and let $g = \mathrm{Max}\, g\, \alpha$. We have

$$\alpha\colon \mathrm{Dom}\, g \le \mathrm{Ran\,Elimseq}(f,A),\ \mathrm{Ln\,Sup}\, g \le \mathrm{Log}(A^9 \cdot (A \# A)),$$
$$\mathrm{Sup\,Sup}\, g \le X_{A+\mathrm{Ln}\, A} + p + \tilde{\exists}$$

and we have $\alpha_g[0]$, so α holds by MAX.

Suppose

$$\exists i(1 \le i \le \mathrm{Ln}\, A' \;\&\; A'(i) \notin \mathrm{Dom}\, g).$$

By BLNP there exists a minimal such i. Suppose $i = 1$. Then $\mathrm{Deg}(f, A'(1)) = 0$. Let $g_1 = g \cup \{\langle A'(1), A'(1)\rangle\}$. Then $\alpha_g[g_1]$, so $g_1 \le g$, a contradiction, and thus $i \neq 1$. Let $z = \mathrm{Newvar}\, A'(i)$, let $a' = \mathrm{Arglast}(f, A'(i))$, and let

$$g_1 = g \cup \{\langle A'(i), \tilde{\exists} z(p{*}a'^{*}{*}z \tilde{\&} g(A'(i-1)))\rangle\}.$$

By (23.60) and (23.59), $\mathrm{Ln}\, g_1(A'(i)) \le \mathrm{Ln}\, g(A'(i-1)) + (8 + \mathrm{Index}\, f)$. Let

$$k = \mathrm{Ln}\, A'(i) + (8 + \mathrm{Index}\, f) \cdot \mathrm{Deg}(f, A'(i)),$$

so that $\mathrm{Ln}\, g_1(A'(i)) \le k$. Since f occurs in $A'(i)$ we have $\mathrm{Index}\, f \le \mathrm{Ln}\, A'(i)$, and $\mathrm{Deg}(f, A'(i)) \le \mathrm{Ln}\, A'(i)$, so that $k \le \mathrm{Log}(A'(i)^9 \cdot (A'(i) \# A'(i)))$. Therefore $\alpha_g[g_1]$, so $g_1 \le g$, a contradiction, and thus $\mathrm{Dom}\, g = \mathrm{Ran}\, A' = \mathrm{Ran\,Elimseq}(f,A)$. Hence $scope_{\exists!g}$ (27.31).

Suppose $scope_{\exists!g}$ $(27.31)_g[g_1]$, and suppose

$$\exists i(1 \le i \le \mathrm{Ln}\, A' \;\&\; g(A'(i)) \neq g_1(A'(i))).$$

By BLNP there exists a minimal such i. Suppose $i = 1$. Then $g(A'(i)) = g_1(A'(i))$, a contradiction, and thus $i \neq 1$. Then again $g(A'(i)) = g_1(A'(i))$, a contradiction, and thus $g = g_1$. Thus *con* (27.31), and thus (27.31). □

In an effort to make the rest of this chapter less unreadable, let us make the following conventions. Any theorem marked * is understood to have the hypothesis

$$\mathrm{extf}_0(t_1, t, f, x', y, y_1, p, C_0', C_1'),$$

called the *tacit hypothesis*, and any nonlogical symbol introduced by a defining axiom marked *, or introduced (by "define ...") in the proof of a theorem marked *, is understood to have as its first arguments t_1, t, f, x', y, y_1, p, C'_0, C'_1. In proving theorems marked *, we will not introduce and discharge the tacit hypothesis, and if (ξ) is such a theorem, by $hyp\,(\xi)$ we mean the hypothesis of the theorem as written (not the tacit hypothesis).

27.32* *Def.* $\text{Tratf}\,A = g \leftrightarrow hyp\,(27.31)\ \&\ scope_{\exists! g}\,(27.31)$, otherwise $g = 1$.

27.33* *Def.* $\text{Trat}\,A = \text{Tratf}\,A(A)$.

This is the arithmetization of A*, for A an atomic formula, in the discussion of function symbols in [Sh,§4.6]. See (i) of that discussion for the following result. The restriction to atomic formulas will be removed shortly.

27.34* *Thm.* A is an atomic formula of $\text{Lang}\,t_1 \to D'_{27.34}(A)$ is a proof in t_1 of $A \tilde{\leftrightarrow} \text{Trat}\,A$.

Proof. Suppose $hyp\,(27.34)$ and define $B' = \text{Elimseq}(f, A)$, $g = \text{Tratf}\,A$. Suppose $1 \le i < \text{Ln}\,B'$ and define $z = \text{Newvar}\,B'(i)$, $a' = \text{Arglast}(f, B'(i))$,

$$C'_2 = \langle B'(i+1) \tilde{\leftrightarrow} g(B'(i+1))\rangle * \langle y \tilde{=} f{*}x'^{*} \tilde{\leftrightarrow} p{*}x'^{*}{*}y\rangle.$$

We want to construct a proof from C'_2 of $B'(i) \tilde{\leftrightarrow} g(B'(i))$. Recall that $g(B'(i)) = \tilde{\exists} z(p{*}a'^{*}{*}z \tilde{\&} g(B'(i+1)))$. Define

$$C'_3 = C'_2 * D'_{25.6}(a' * \langle z\rangle, x' * \langle y\rangle, y \tilde{=} f{*}x'^{*} \tilde{\leftrightarrow} p{*}x'^{*}{*}y).$$

By the substitution rule (25.6), $C'_3\colon C'_2 \tilde{\vdash} z \tilde{=} f{*}a'^{*} \tilde{\leftrightarrow} p{*}a'^{*}{*}z$. Define

$$C'_4 = C'_3 * D'_{25.34}(\tilde{\exists} z(z \tilde{=} f{*}a'^{*} \tilde{\&} B'(i+1)), g(B'(i)), \langle z \tilde{=} f{*}a'^{*}\rangle * \langle B'(i+1)\rangle, \langle p{*}a'^{*}{*}z\rangle * \langle g(B'(i+1))\rangle).$$

By the equivalence theorem (25.34),

$$C'_4\colon C'_2 \tilde{\vdash} \tilde{\exists} z(z \tilde{=} f{*}a'^{*} \tilde{\&} B'(i+1) \tilde{\leftrightarrow} g(B'(i)))).$$

Define

$$C'_5 = C'_4 * D'_{25.58}(f{*}a'^{*}, z, B'(i+1)) * \langle B'(i+1)_z[f{*}a'^{*}] \tilde{\leftrightarrow} g(B'(i+1))\rangle.$$

By the corollary to the equality theorem (25.58), $C'_5\colon C'_2 \tilde{\vdash} C'_5(\text{Ln}\,C'_5)$, but $B'(i+1)_z[f{*}a'^{*}] = B'(i)$. Thus $1 \le i < \text{Ln}\,B' \to C'_5\colon C'_2 \tilde{\vdash} B'(i) \tilde{\leftrightarrow} g(B'(i))$. Define

$$D' = \text{Reverse}\{\langle i, C'_5\rangle : i \in \text{Dom}\,B'[1, \text{Ln}\,B' - 1]\}.$$

Then D': $\langle y \mathrel{\hat{=}} f{*}x'^{*} \rightleftarrows p{*}x'^{*}{*}y \rangle \mathrel{\tilde{\vdash}} A \rightleftarrows g(A)$, and thus (27.34).

27.35* *Def.* $\operatorname{Trfn} A = \operatorname{Trext}(\{\langle B, \operatorname{Trat} B\rangle : B \in \operatorname{Atoms} A\}, A)$.

27.36* *Def.* $\operatorname{Tr} A = \operatorname{Trfn} A(A)$.

27.37* *Thm.* A is a formula of $\operatorname{Lang} t_1 \to \operatorname{Tr} A$ is a formula of $\operatorname{Lang} t$ & (A is a formula of $\operatorname{Lang} t \to \operatorname{Tr} A = A$).

Proof. Suppose $\exists A \neg(27.37)$. By BLNP there exists a minimal such A. Then A is an atomic formula, so that $\operatorname{Tr} A = \operatorname{Trat} A$. But by (27.31) this is impossible. Thus (27.37).

27.38* *Thm.* A is a formula of $\operatorname{Lang} t_1 \to D'_{27.38}(A)$ is a proof in t_1 of $A \rightleftarrows \operatorname{Tr} A$.

Proof. Define

$$B' = \operatorname{Enum} \operatorname{Atoms} A,$$
$$B'_1 = \{\langle i, \operatorname{Trat} B'(i)\rangle : i \in \operatorname{Dom} B'\},$$
$$D' = \{\langle i, D'_{27\,34}(B'(i))\rangle : i \in \operatorname{Dom} B'\}^{*} * D'_{25\,34}(A, \operatorname{Tr} A, B', B'_1).$$

Then (27.38) by (27.34) and the equivalence theorem (25.34), which is applicable by (27.10). □

The next task is to give a predicative arithmetization of the theorem that an extension by definition of a function symbol is conservative. It is a semantic triviality that such an extension does not alter the validity of any formula, but the argument by Gödel's Completeness Theorem that it is a conservative extension yields no clue as to how much longer is a proof with the new function symbol eliminated. We cannot follow the proof of (ii) in the discussion of function symbols in [Sh,§4.6] because it is based on the Theorem on Functional Extensions (in which the uniqueness condition is not assumed), whose proof in [Sh,§4.5] uses Herbrand's Theorem, which in turn depends on the Consistency Theorem; this argument yields a super-exponential bound. (I do not know whether the Theorem on Functional Extensions can be established predicatively; we have not used such extensions in the development of our theory.) The difficulty in a direct proof, similar to the one for an extension by definition of a predicate symbol, is that the translation of a substitution axiom is not in general a substitution axiom—the analogue of (27.19) fails because the term a may contain the new function symbol f. There is a direct proof in [Kl,§74], but it is very intricate. Here is a different proof; so far as I know, it is new.

The idea can be described very quickly. Consider an extension T_1 of T by definition of a function symbol f. For any variable-free term a of T_1 we define a variable-free term a^c of T_c (this is the theory obtained by adjoining all special constants and special axioms of T; see [Sh,§4.2]) by replacing each part of a of the form $fa_1 \ldots a_\nu$ by the corresponding special constant, working from right to left. By the existence and uniqueness conditions and the special axioms, $A_x[a]^*$ and $A^*_x[a^c]$ are equivalent in T_c. Therefore if C is a closed substitution axiom of T_1, then C^* is equivalent in T_c to a substitution axiom, and since T_c is a conservative extension of T, C^* is a theorem of T. If C is an arbitrary substitution axiom of T_1, then $\vdash_T C^*$ by the Theorem on Constants.

Now let us examine the argument in full detail. Let T be a theory, let p be a $(\nu + 1)$-ary predicate symbol such that

$$\vdash_T \exists y\, px_1 \ldots x_\nu y$$

(existence condition) and

$$\vdash_T px_1 \ldots x_\nu y \;\&\; px_1 \ldots x_\nu y_1 \rightarrow y = y_1$$

(uniqueness condition), and let T_1 be the theory obtained by adjoining to T a new ν-ary function symbol f and the defining axiom

$$y = fx_1 \ldots x_\nu \leftrightarrow px_1 \ldots x_\nu y.$$

Let A be an atomic formula of T_{1c}; we define A^* by induction on the number of occurrences of f in A. If there are none, then A^* is A. Otherwise, let B be the atomic formula with one less occurrence of f obtained by replacing the rightmost occurrence in A of a term beginning with f by z, where z is the first variable in alphabetical order not occurring in A, so that A is $B_z[fa_1 \ldots a_\nu]$ where $a_1, \ldots, a_\nu$ do not contain f. Then A^* is

$$\exists z(pa_1 \ldots a_\nu z \;\&\; B^*).$$

For a non-atomic formula A of T_{1c}, we define A^* by replacing each atomic part C of A by C^*. This is the construction of [Sh,§4.6], which has been arithmetized in (27.31).

Let us show that if a is a term of T_{1c} with no occurrences of f, then

i. $A_x[a]^*$ is $A^*_x[a]$.

It suffices to prove (i) for A atomic. Then the proof is by induction on the number of occurrences of f in A. If there are none, then both formulas are $A_x[a]$. Otherwise, using the notation introduced above, we see that the following are all the same formula:

$$A_x[a]^*,$$

$$B_z[fa_1 \ldots a_\nu]_x[a]^*,$$

$$B_x[a]_z[fa_{1x}[a] \ldots a_{\nu x}[a]]^*,$$

$$\exists z(pa_{1x}[a] \ldots a_{\nu x}[a]z \; \& \; B_x[a]^*),$$

$$\exists z(pa_{1x}[a] \ldots a_{\nu x}[a]z \; \& \; B^*_x[a]),$$

$$A^*_x[a].$$

The induction hypothesis was used in the next to last step.

Let a be a variable-free term of T_{1c}. We define a variable-free term a^c of T_c by induction on the number of occurrences of f in a. If there are none, then a^c is a. Otherwise, let b be the term with one less occurrence of f obtained by replacing the rightmost occurrence in a of a term beginning with f by y, so that a is $b_y[fa_1 \ldots a_\nu]$ where $a_1 \ldots a_\nu$ do not contain f. Let r be the special constant for $\exists y pa_1 \ldots a_\nu y$; then a^c is $b_y[r]^c$.

Now let us show that if a is a variable-free term of T_{1c}, then

ii. $\vdash_{T_c} A_x[a]^* \leftrightarrow A^*_x[a^c]$.

By the Equivalence Theorem, it suffices to prove (ii) for A atomic. We can assume that there is precisely one free occurrence of x in A—if there are none, we have (ii) by a tautology, and otherwise we can write $A_x[a]$ in the form $A'_{x_1 \ldots x_\lambda}[a \ldots a]$ and obtain (ii) by induction on λ from the case $\lambda = 1$. Now the proof is by induction on the number of occurrences of f in $A_x[a]$. If there are none, then we have (ii) by a tautology. Otherwise, use the notation introduced above. If the rightmost occurrence of a term beginning with f is not in the occurrence of a that has been substituted for x, then $A_x[a]$ is $B_x[a]_z[fa_1 \ldots a_\nu]$, so that $A_x[a]^*$ is $\exists z(pa_1 \ldots a_\nu z \; \& \; B_x[a]^*)$, which by the induction hypothesis is equivalent in T_c to $\exists z(pa_1 \ldots a_\nu z \; \& \; B^*_x[a^c])$; i.e., to $A^*_x[a^c]$. Otherwise $A_x[a]$ is $A_x[b_y[fa_1 \ldots a_\nu]]$, so

$$A_x[a]^* \text{ is } \exists z(pa_1 \ldots a_\nu z \;\&\; A_x[b']^*)$$

where b' is $b_y[z]$. Now

$$\vdash_{T_c} \exists y pa_1 \ldots a_\nu y$$

by the existence condition and the Substitution Rule, so

$$\text{iii. } \vdash_{T_c} pa_1 \ldots a_\nu r$$

by the special axiom for r. Also,

$$\vdash_{T_c} pa_1 \ldots a_\nu r \;\&\; pa_1 \ldots a_\nu z \to r = z$$

by the uniqueness condition and the Substitution Rule, so

$$\vdash_{T_c} pa_1 \ldots a_\nu z \to r = z$$

and a fortiori

$$\vdash_{T_c} pa_1 \ldots a_\nu z \;\&\; A_x[b']^* \to r = z.$$

But

$$\vdash_{T_c} r = z \to (A_x[b']^*_z[r] \leftrightarrow A_x[b']^*)$$

by the Equality Theorem, so

$$\vdash_{T_c} pa_1 \ldots a_\nu z \;\&\; A_x[b']^* \to A_x[b']^*_z[r]$$

by a tautology. By $\exists$-introduction,

$$\text{iv. } \vdash_{T_c} \exists z(pa_1 \ldots a_\nu z \;\&\; A_x[b']^*_z[r]) \to A_x[b']^*_z[r].$$

Conversely, since

$$\vdash_{T_c} pa_1 \ldots a_\nu r \;\&\; A_x[b']^*_z[r] \to \exists z(pa_1 \ldots a_\nu z \;\&\; A_x[b']^*)$$

by a Substitution Axiom, we have by (iii) that

$$\text{v. } \vdash_{T_c} A_x[b']^*_z[r] \to \exists z(pa_1 \ldots a_\nu z \;\&\; A_x[b']^*).$$

By (iv) and (v), $A_x[a]^*$ is equivalent in T_c to $A_x[b']^*_z[r]$, which by (i) is $A_x[b'_z[r]]^*$, i.e. $A_x[b_y[r]]^*$. By the induction hypothesis, this is equivalent in T_c to $A^*_x[b_y[r]^c]$; i.e., to $A^*_x[a^c]$. This proves (ii).

Now we are ready to prove that if $\vdash_{T_1} A$ then $\vdash_T A^*$. The proof is by induction on theorems (see [Sh,§3.1]). Tautological consequence and $\exists$-introduction present no problem, by the way in which we defined A^* for A non-atomic (to borrow from our arithmetical terminology, the mapping $A \mapsto A^*$ is a translation function), so we need only show that if A is an axiom of T_1 then A^* is a theorem of T. Let T_0 be the theory obtained by adjoining new constants $e_1, e_2, e_3, \ldots$ (distinct from f), but no new nonlogical axioms, to T. Let A^0 be $A_{w_1 \ldots w_\mu}[e_1 \ldots e_\mu]$ where $w_1, \ldots, w_\mu$ are in alphabetical order the free variables of A. Notice that A^{0*} is A^{*0}, by (i), so by the Theorem on Constants it suffices to show that $\vdash_{T_0} A^{0*}$ for A an axiom of T_1.

Let A be an axiom of T. Then A^{0*} is an instance of A.

Let A be a substitution axiom of T_1. Then A^0 is a closed substitution axiom $B_x[a] \to \exists x B$, and by (ii), $B_x[a]^*$ is equivalent in T_{0c} to $B^*_x[a^c]$, so that A^{0*} is equivalent in T_{0c} to the substitution axiom $B^*_x[a^c] \to \exists x B^*$. But A^{0*} is a formula of T_0 and T_{0c} is a conservative extension of T_0, so A^{0*} is a theorem of T_0.

Let A be an equality axiom for f. Then A^0 is

$$a_1 = b_1 \to \cdots \to a_\nu = b_\nu \to fa_1 \ldots a_\nu = fb_1 \ldots b_\nu$$

where $a_1, \ldots, b_\nu$ are $e_1, \ldots, e_{2\nu}$ in some order. Let r be the special constant for $\exists y p a_1 \ldots a_\nu y$ and let s be the special constant for $\exists y p b_1 \ldots b_\nu y$. By (ii), A^{0*} is equivalent in T_{0c} to

$$a_1 = b_1 \to \cdots \to a_\nu = b_\nu \to r = s.$$

But

$$\vdash_{T_{0c}} a_1 = b_1 \to \cdots \to a_\nu = b_\nu \to (pa_1 \ldots a_\nu s \leftrightarrow pb_1 \ldots b_\nu s)$$

by the Equality Theorem, and

$$\vdash_{T_{0c}} pb_1 \ldots b_\nu s$$

by the special axiom for s and the existence condition, so A^{0*} is a theorem of T_{0c} and hence of T_0.

Finally, let A be the defining axiom of f. Then A^0 is

$$\mathrm{b} = \mathrm{f}\mathrm{a}_1 \ldots \mathrm{a}_\nu \leftrightarrow \mathrm{p}\mathrm{a}_1 \ldots \mathrm{a}_\nu \mathrm{b}$$

where $\mathrm{b}, \mathrm{a}_1, \ldots, \mathrm{a}_\nu$ are $\mathrm{e}_1, \ldots, \mathrm{e}_{\nu+1}$ in some order. Again let r be the special constant for $\exists y \mathrm{p}\mathrm{a}_1 \ldots \mathrm{a}_\nu y$. By (ii), A^{0*} is equivalent in T_{0c} to

$$\mathrm{b} = \mathrm{r} \leftrightarrow \mathrm{p}\mathrm{a}_1 \ldots \mathrm{a}_\nu \mathrm{b},$$

so that A^{0*} is a theorem of T_{0c}, and hence of T_0, by the existence and uniqueness conditions and the special axiom for r.

Since A^* is A if A is a formula of T, this shows that T_1 is a conservative extension of T.

Now it is a straightforward task to give a predicative arithmetization of this argument.

27.39* *Thm.* D_0' is a proof in t_1 of $A \to D_{27\,39}'(D_0')$ is a proof in t of $\mathrm{Tr}\, A$.

The proof and the construction of $D_{27\,39}'$ are omitted. More precisely, I have omitted to write them down. In a previous version of this chapter I carried out a detailed arithmetization. But as Sam Buss pointed out, there was a conceptual error in the original metamathematical argument, and this was simply copied in the arithmetization. The moral is the familiar one that one's capacity for self-deception is great. Perhaps this can be avoided by submitting oneself to the discipline of a completely formalized proof, but this is an ideal that is seldom achieved in mathematics, and certainly not in the present work.

Now we remove the restriction that the right hand side of the defining axiom be an atomic formula.

27.40 *Def.* $\mathrm{extf}(t_1, t, f, x', y, y_1, D, C_0', C_1') \leftrightarrow t_1$ and t are theories & f is a function symbol & $f \notin \mathrm{Lang}\, t$ & $x' * \langle y \rangle * \langle y_1 \rangle$ is an injective sequence of variables & $\mathrm{Index}\, f = \mathrm{Ln}\, x'$ & D is a formula of $\mathrm{Lang}\, t$ & $\mathrm{Ran}\,\mathrm{Free}\, D \subseteq \mathrm{Ran}(x' * \langle y \rangle)$ & C_0' is a proof in t of $\tilde{\exists} y D$ & y_1 is substitutable for y in D & C_1' is a proof in t of $D \tilde{\&} D_y[y_1] \tilde{\to} y \tilde{=} y_1$ & $t_1 = t[y \tilde{=} f * x'^* \tilde{\leftrightarrow} D]$.

27.41 *Def.* $\mathrm{Tr}_1(t_1, t, f, x', y, y_1, D, C_0', C_1', u, A) =$
$\mathrm{Trp}(\mathrm{P}_{\mathrm{Ln}\, x'+1, t}, x' * \langle y \rangle, D, u, \mathrm{Tr}\,(t_1, t, f, x', y, y_1, \mathrm{P}_{\mathrm{Ln}\, x'+1, t}, C_0', C_1', A))$.

27.42 *Thm.* $\mathrm{extf}(t_1, t, f, x', y, y_1, D, C_0', C_1')$ & D_0' is a proof in t of $A \to$ $D_{27\,42}'(t_1, t, f, x', y, y_1, D, C_0', C_1', D_0')$ is a proof in t of
$\mathrm{Tr}_1(t_1, t, f, x', y, y_1, D, C_0', C_1', u_{27\,42}(t_1, t, f, x', y, y_1, D, C_0', C_1', D_0'), A)$.

Proof. Suppose *hyp* (27.42) and define

$$
\begin{aligned}
&p = \mathrm{P}_{\mathrm{Ln}\,x'+1,t}\ , \\
&t_2 = t[\langle p{*}x'^{*}{*}y \stackrel{\sim}{\leftrightarrow} D\rangle], \\
&t_3 = t_2[\langle y \stackrel{\sim}{=} f{*}x'^{*} \stackrel{\sim}{\leftrightarrow} p{*}x'^{*}{*}y\rangle], \\
&C'_2 = C'_0 * C'_{25.32}(p{*}x'^{*}{*}y, D, y) * \langle \tilde{\exists} y p{*}x'^{*}{*}y\rangle, \\
&C'_3 = C'_1 * D'_{25\,6}(\langle y_1\rangle, \langle y\rangle, p{*}x'^{*}{*}y \stackrel{\sim}{\leftrightarrow} D) * \langle p{*}x'^{*}{*}y_1 \stackrel{\sim}{\leftrightarrow} D_y[y_1]\rangle *
\end{aligned}
$$
$$\langle p{*}x'^{*}{*}y \,\tilde{\&}\, p{*}x'^{*}{*}y_1 \stackrel{\sim}{\rightarrow} y \stackrel{\sim}{=} y_1\rangle.$$

Then $\mathrm{extf}_0(t_3, t_2, f, x', y, y_1, p, C'_2, C'_3)$. Define

$$
\begin{aligned}
&D'_1 = D'_{27\,39}(t_3, t_2, f, x', y, y_1, p, C'_2, C'_3, D'_0), \\
&u = D'^{\,*}_1, \\
&D' = D'_{27\,30}(p, x', D, D'_1).
\end{aligned}
$$

Then *con* (27.42), and thus (27.42).

Chapter 28

Interpretations

The proof in [Sh,§4.7] of the Interpretation Theorem has a straightforward predicative arithmetization. We begin with the notion of an interpretation i, with universe u_0, of one language in another.

28.1 *Def.* $\text{interp}_0(i, u_0, l_1, l_2) \leftrightarrow l_1$ and l_2 are languages & u_0 is a predicate symbol & $\text{Index}\, u_0 = 1$ & $u_0 \in l_2$ & $\text{Dom}\, i = l_1$ & $\text{Ran}\, i \subseteq l_2$ & $\forall v(v \in l_1 \rightarrow \text{Index}\, i(v) = \text{Index}\, v$ & $(i(v)$ is a predicate symbol $\leftrightarrow v$ is a predicate symbol$))$.

28.2 *Def.* $u_{(i)} = \{\langle j, v\rangle : j \in \text{Dom}\, u$ & $(u(j) \in \text{Dom}\, i \rightarrow v = i(u(j)))$ & $(u(j) \notin \text{Dom}\, i \rightarrow v = u(j))\}$.

This construction is used to interpret a term or an atomic formula u: replace each nonlogical symbol by i of it. We construct the interpretation $A^{(i,u_0)}$ of a general formula A in two steps.

28.3 *Thm.* $\exists! g(g$ is a function & $\text{Dom}\, g = \text{Formulas}\, A$ & $\forall B \forall C \forall x(B \in \text{Dom}\, g$ & $C \in \text{Dom}\, g$ & x is a variable $\rightarrow (1)$ & $\cdots$ & $(5)))$, *where*

1. $\text{Ln}\, g(B) \leq 8 \cdot \text{Ln}\, B \leq \text{Log}\, B^8$ & $\text{Sup}\, g(B) \leq \text{Max}(\text{Sup}\, B, \text{Sup}\, i) + \bar{\exists} + \bar{\vee} + \bar{\neg} + u_0$,
2. $B \in \text{Atoms}\, A \rightarrow g(B) = B_{(i)}$,
3. $\tilde{\neg} B \in \text{Dom}\, g \rightarrow g(\tilde{\neg} B) = \tilde{\neg} g(B)$,
4. $B \tilde{\vee} C \in \text{Dom}\, g \rightarrow g(B \tilde{\vee} C) = g(B) \tilde{\vee} g(C)$,
5. $\tilde{\exists} x B \in \text{Dom}\, g \rightarrow g(\tilde{\exists} x B) = \tilde{\exists} x(u_0 * x \tilde{\&} g(B))$.

Proof. Write α as an abbreviation for $scope_{\exists! g}$ (28.3) but with $\text{Dom}\, g =$

Formulas A replaced by $\mathrm{Dom}\, g \subseteq \mathrm{Formulas}\, A$ & $\forall B(B \in \mathrm{Dom}\, g \rightarrow \mathrm{Formulas}\, B \subseteq \mathrm{Dom}\, g)$, and let $g = \mathrm{Max}\, g\, \alpha$. We have $\alpha_g[0]$, so α holds by MAX. Suppose

$$\exists B_0(B_0 \in \mathrm{Formulas}\, A \;\&\; B_0 \notin \mathrm{Dom}\, g).$$

By BLNP there exists a minimal such B_0. Clearly $B_0 \notin \mathrm{Atoms}\, A$, so there exist B, C, and x such that B and C are formulas & x is a variable & $(B_0 = \tilde{\neg} B \vee B_0 = B \tilde{\vee} C \vee B_0 = \tilde{\exists} x B)$, but this is impossible. Thus $\mathrm{Dom}\, g = \mathrm{Formulas}\, A$ and so $scope_{\exists^! g}(28.3)$.

Suppose $scope_{\exists^! g}(28.3)_g[g_1]$, and suppose

$$\exists B_0(B_0 \in \mathrm{Formulas}\, A \;\&\; g(B_0) \neq g_1(B_0)).$$

By BLNP there exists a minimal such B_0, which is impossible. Thus $g = g_1$, and thus (28.3).

28.4 *Def.* $A_{(i,u_0)} = (\mathrm{Min}\, g\, scope_{\exists^! g}(28.3))(A)$.

28.5 *Def.* $\mathrm{Univ}(u_0, x') = \{\langle j, u_0 {*} x'(j)\rangle : j \in \mathrm{Dom}\, x'\}$.

28.6 *Def.* $A^{(i,u_0)} = \mathrm{Impl}(\mathrm{Univ}(u_0, \mathrm{Free}\, A) * \langle A_{(i,u_0)}\rangle)$.

Now we can define an interpretation of one theory in another. Again, proofs are included as part of the structure in order to obtain a bounded predicate symbol.

28.7 *Def.* $\mathrm{interp}(i, u_0, C', x'', C_0'', C_1'', t_1, t_2) \leftrightarrow t_1$ and t_2 are theories & $\mathrm{interp}_0(i, u_0, \mathrm{Lang}\, t_1, \mathrm{Lang}\, t_2)$ & C' is a proof in t_2 of $\tilde{\exists} X_0(u_0 {*} X_0)$ & $\mathrm{Dom}\, x'' = \mathrm{Dom}\, C_0'' = \{f \in \mathrm{Lang}\, t_1 : f$ is a function symbol$\}$ & $\mathrm{Dom}\, C_1'' = \mathrm{Ax}\, t_1$ & (1) & (2), *where*

1. $\forall f(f \in \mathrm{Dom}\, x'' \rightarrow x''(f)$ is an injective sequence of variables & $\mathrm{Ln}\, x''(f) = \mathrm{Index}\, f$ & $c_0''(f)$ is a proof in t_2 of $\mathrm{Impl}(\mathrm{Univ}(u_0, x''(f)) * \langle u_0 {*} i(f) {*} x''(f)^{*}\rangle))$,
2. $\forall A(A \in \mathrm{Ax}\, t_1 \rightarrow C_1''(A)$ is a proof in t_2 of $A^{(i,u_0)})$.

Given an interpretation of one theory in another, we want a bounded function symbol that converts a proof of a theorem in the one theory into a proof of its interpretation in the other theory. We need two preliminary results.

28.8 *Thm.* $\mathrm{interp}(i, u_0, C', x'', C_0'', C_1'', t_1, t_2)$ & a is a term of $\mathrm{Lang}\, t_1 \rightarrow D'_{28.8}(i, u_0, x'', C_0'', a)$ is a proof in t_2 of $\mathrm{Impl}(\mathrm{Univ}(u_0, \mathrm{Free}\, a) * \langle u_0 {*} a_{(\imath)} \rangle)$.

Proof. Suppose *hyp* (28.8) and define

$r_0 = \{\langle b, \mathrm{Ln}\, b\rangle : b \in \mathrm{Terms}\, a\}$,
$a' = \mathrm{Enumer}(\mathrm{Terms}\, a, r_0)$,
$B' = \{\langle j, \mathrm{Impl}(\mathrm{Univ}(u_0, \mathrm{Free}\, a'(j)) * \langle u_0 {*} a'(j)_{(\imath)} \rangle)\rangle : j \in \mathrm{Dom}\, a'\}$.

Suppose $1 \leq j \leq \mathrm{Ln}\, a'$. We distinguish two cases: 1. $a'(j)$ is a variable and 2. $\neg(a'(j)$ is a variable). Define

$D_1' = u_0 {*} a'(j) \eqsim u_0 {*} a'(j)$,
$f = \langle a'(j)(1)\rangle$,
$b' = \mathrm{Arg}\, a'(j)$,
$c' = \{\langle k, b'(k)_{(\imath)}\rangle : k \in \mathrm{Dom}\, b'\}$,
$D_2' = C_0''(f) * D'_{25.6}(c', x''(f), \mathrm{Impl}(\mathrm{Univ}(u_0, x''(f)) * \langle u_0 {*} i(f) {*} x''(f)^{*}\rangle)) * \langle B'(j)\rangle$,
$D_3' = D_\mu'$ in case (μ), for $1 \leq \mu \leq 2$.

Thus $1 \leq j \leq \mathrm{Ln}\, a' \rightarrow D_3'$ is a proof in $t_2[B'[1, j-1]]$ of $B'(j)$. Define

$D' = \{\langle j, D_3'\rangle : j \in \mathrm{Dom}\, B'\}^{*}$.

Then *con* (28.8), and thus (28.8).

28.9 *Thm.* $\mathrm{interp}(i, u_0, C', x'', C_0'', C_1'', t_1, t_2)$ & x' is a sequence of variables & A is a formula of $\mathrm{Lang}\, t_1$ & $\mathrm{Ran\, Free}\, A \subseteq \mathrm{Ran}\, x'$ & D_0' is a proof in t_2 of $\mathrm{Impl}(\mathrm{Univ}(u_0, x') * \langle A_{(\imath, u_0)}\rangle) \rightarrow D'_{28.9}(i, u_0, C', x'', C_0'', x', A, D_0')$ is a proof t_2 of $A^{(\imath, u_0)}$.

Proof. Suppose *hyp* (28.9) and define

$y' = \mathrm{Enum}\{y \in \mathrm{Ran}\, x' : y \notin \mathrm{Ran\, Free}\, A\}$,
$B' = \{\langle j, \mathrm{Impl}(\mathrm{Univ}(u_0, y'[j, \mathrm{Ln}\, y']) * \langle A^{(\imath, u_0)}\rangle)\rangle : j \in \mathrm{Dom}\, y' \cup \{\mathrm{Ln}\, y' + 1\}\}$.

Suppose $1 \leq j \leq \mathrm{Ln}\, y'$ and define

$D_1' = \langle B'(j)\rangle * \langle \tilde{\exists} y'(j)(u_0 {*} y'(j) \eqsim B'(j+1))\rangle * C' * \langle \tilde{\exists} X_0(u_0 {*} X_0)\rangle * D'_{25.35}(u_0 {*} X_0, X_0, y'(j)) * \langle \tilde{\exists} y'(j)(u_0 {*} y'(j))\rangle * \langle B'(j+1)\rangle$.

Thus $1 \leq j \leq \mathrm{Ln}\, y' \rightarrow D_1'$ is a proof in $t_2[\langle B'(j)\rangle]$ of $B'(j+1)$. Define

$D' = D_0' * \langle B'(1)\rangle * \{\langle j, D_1'\rangle : j \in \mathrm{Dom}\, y'\}^{*}$.

Then *con* (28.9), and thus (28.9).

28.10 *Thm.* (*interpretation theorem*) $\text{interp}(i, u_0, C', x'', C_0'', C_1'', t_1, t_2)$ & D_0' is a proof in t_1 of $A \rightarrow D_{28\,10}'(i, u_0, C', x'', C_0'', C_1'', D_0')$ is a proof in t_2 of $A^{(i,u_0)}$.

Proof. Suppose *hyp* (28.10) and suppose $1 \leq j \leq \text{Ln}\, D_0'$. We distinguish five cases:

1. $\exists B \exists x \exists a (a$ is substitutable for x in B & $D_0'(j) = B_x\lceil a \rceil \tilde{\rightarrow} \tilde{\exists} x B)$,
2. $D_0'(j)$ is an identity axiom $\vee$ $D_0'(j)$ is an equality axiom,
3. $D_0'(j) \in \text{Ax}\, t_1$,
4. $D_0'(j)$ is a tautological consequence of $D_0'[1, j-1]$,
5. $\exists k (1 \leq k < j$ & $D_0'(j)$ can be inferred from $D_0'(k)$ by $\exists$-introduction).

Define B, x, and a as in (1), and define

$y' = \text{Enum}\{y \in \text{Ran Free}\,(B * a) : y \neq x\}$,
$D_{10}' = D_{28\,8}'(i, u_0, x'', C_0'', a) * \langle \text{Impl}(\text{Univ}(u_0, y') * \langle u_0 {*} a_{(i)} \rangle) \rangle *$
$\langle u_0 {*} a_{(i)} \tilde{\&} (B_{(i,u_0)})_x \lceil a_{(i)} \rceil \tilde{\rightarrow} \tilde{\exists} x (u_0 {*} x \tilde{\&} B_{(i,u_0)}) \rangle *$
$\langle \text{Impl}(\text{Univ}(u_0, y') * \langle (B_{(i,u_0)})_x \lceil a_{(i)} \rceil \rangle * \langle \tilde{\exists} x (u_0 {*} x \tilde{\&} B_{(i,u_0)}) \rangle) \rangle *$
$\langle \text{Impl}(\text{Univ}(u_0, y') * \langle D_0'(j)_{(i,u_0)} \rangle) \rangle$,
$D_1' = D_{28\,9}'(i, u_0, C', x'', C_0'', x', D_0'(j), D_{10}')$,
$D_2' = \langle D_0'(j)_{(i,u_0)} \rangle * \langle D_0'(j)^{(i,u_0)} \rangle$,
$D_3' = C_1''(D_0'(j))$,
$x' = \text{Free}\, D_0'[1, j]^*$,
$D_{40}' = \{\langle k, D_0'(k)^{(i,u_0)} \rangle : k \in \text{Dom}\, D_0'[1, j-1]\} *$
$\langle \text{Impl}(\text{Univ}(u_0, x') * \langle D_0'(j)_{(i,u_0)} \rangle) \rangle$,
$D_4' = D_{28\,9}'(i, u_0, C', x'', C_0'', x', D_0'(j), D_{40}')$.

Define k as in (5), define x_0, C, and D as in

$\exists x_0 \exists C \exists D (x_0$ is a variable & C and D are formulas & $\neg(x_0$ is free in $D)$ & $D_0(k) = C \tilde{\rightarrow} D$ & $D_0'(j) = \tilde{\exists} x_0 C \tilde{\rightarrow} D)$,

and define

$x_0' = \text{Free}\, D_0'(j)$,
$D_5' = \langle \text{Impl}(\text{Univ}(u_0, x_0') * \langle u_0 {*} x_0 \rangle * \langle C_{(i,u_0)} \rangle * \langle D_{(i,u_0)} \rangle) \rangle *$
$\langle u_0 {*} x_0 \tilde{\&} C_{(i,u_0)} \tilde{\rightarrow} \text{Impl}(\text{Univ}(u_0, x_0') * \langle D_{(i,u_0)} \rangle) \rangle *$
$\langle \tilde{\exists} x_0 (u_0 {*} x_0 \tilde{\&} C_{(i,u_0)}) \tilde{\rightarrow} \text{Impl}(\text{Univ}(u_0, x_0') * \langle D_{(i,u_0)} \rangle) \rangle * \langle D_0'(j)^{(i,u_0)} \rangle$,

$D'_6 = D'_\mu$ in case (μ), for $1 \leq \mu \leq 5$.

Thus $1 \leq j \leq \operatorname{Ln} D'_0 \to D'_6$ is a proof in

$$t_2[\operatorname{Ran}\{\langle l, D'_0(l)^{(\imath, u_0)}\rangle : l \in \operatorname{Dom} D'_0[1, j-1]\}]$$

of $D'_0(j)$. Define

$$D' = \{\langle j, D'_6\rangle : j \in \operatorname{Dom} D'_0\}^*.$$

Then *con* (28.10), and thus (28.10).

Chapter 29

The arithmetization of arithmetic

Now we can begin to prove results about predicative arithmetic within predicative arithmetic. In this chapter we will arithmetize Robinson's theory and show it to be tautologically consistent (and we do this within a theory that is interpretable in Robinson's theory).

29.1 *Def.* $\tilde{0} = F_{0,0}$.

29.2 *Def.* $\overline{\mathrm{S}} = F_{1,0}$.

29.3 *Def.* $\overline{\mathrm{P}} = F_{1,1}$.

29.4 *Def.* $\overline{+} = F_{2,0}$.

29.5 *Def.* $\bar{\cdot} = F_{2,1}$.

29.6 *Def.* $\tilde{\mathrm{S}}a = \overline{\mathrm{S}} * a$.

29.7 *Def.* $\tilde{\mathrm{P}}a = \overline{\mathrm{P}} * a$.

29.8 *Def.* $a \tilde{+} b = \overline{+} * a * b$.

29.9 *Def.* $a \tilde{\cdot} b = \bar{\cdot} * a * b$.

If x is the νth variable in alphabetical order, we use $\tilde{\mathrm{x}}$ as an abbreviation for $X_{\overline{\nu}}$. (Recall that $\overline{\nu}$ is $\mathrm{S} \cdots \mathrm{S}0$ with ν occurrences of S.) Also, we use $\{\mathrm{a}_1, \ldots, \mathrm{a}_\nu\}$ as an abbreviation for $\{\mathrm{a}_1\} \cup \cdots \cup \{\mathrm{a}_\nu\}$.

29.10 *Def.* $\overline{Q}_0 = \langle\{\tilde{0}, \overline{\mathrm{S}}, \overline{\mathrm{P}}, \tilde{+}, \tilde{\cdot}\}, \{\tilde{\mathrm{S}}\tilde{x}\tilde{\neq}\tilde{0},\ \tilde{\mathrm{S}}\tilde{x}\tilde{\simeq}\tilde{\mathrm{S}}\tilde{y}\tilde{\rightarrow}\tilde{x}\tilde{\simeq}\tilde{y},\ \tilde{x}\tilde{+}\tilde{0}\tilde{\simeq}\tilde{x},$ $\tilde{x}\tilde{+}\tilde{\mathrm{S}}\tilde{y}\tilde{\simeq}\tilde{\mathrm{S}}(\tilde{x}\tilde{+}\tilde{y}),\ \tilde{x}\tilde{\cdot}\tilde{0}\tilde{\simeq}\tilde{0},\ \tilde{x}\tilde{\cdot}\tilde{\mathrm{S}}\tilde{y}\tilde{\simeq}\tilde{x}\tilde{\cdot}\tilde{y}\tilde{+}\tilde{x},\ \tilde{\mathrm{P}}\tilde{x}\tilde{\simeq}\tilde{y}\tilde{\leftrightarrow}\tilde{\mathrm{S}}\tilde{y}\tilde{\simeq}\tilde{x}\tilde{\vee}(\tilde{x}\tilde{\simeq}\tilde{0}\tilde{\&}\tilde{y}\tilde{\simeq}\tilde{0})\}\rangle$.

29.11 *Def.* a is a variable-free term of $l \leftrightarrow a$ is variable-free & a is a term of l.

29.12 *Thm.* a is a variable-free term of Lang $\overline{Q}_0 \rightarrow \exists! f(f$ is a function & $\operatorname{Dom} f = \operatorname{Terms} a$ & $\forall b \forall c(b \in \operatorname{Dom} f$ & $c \in \operatorname{Dom} f \rightarrow (1)$ & $\cdots$ & $(6)))$, *where*

1. $f(b) \leq \operatorname{Expln}(2, b) \leq b$,
2. $f(\tilde{0}) = 0$,
3. $\tilde{\mathrm{S}}b \in \operatorname{Dom} f \rightarrow f(\tilde{\mathrm{S}}b) = \mathrm{S}f(b)$,
4. $\tilde{\mathrm{P}}b \in \operatorname{Dom} f \rightarrow f(\tilde{\mathrm{P}}b) = \mathrm{P}f(b)$,
5. $b\tilde{+}c \in \operatorname{Dom} f \rightarrow f(b\tilde{+}c) = f(b) + f(c)$,
6. $b\tilde{\cdot}c \in \operatorname{Dom} f \rightarrow f(b\tilde{\cdot}c) = f(b) \cdot f(c)$.

Proof. Suppose *hyp* (29.12) and write α as an abbreviation for

$$\mathit{scope}_{\exists! f}\,(29.12)$$

but with $\operatorname{Dom} f = \operatorname{Terms} a$ replaced by

$$\operatorname{Dom} f \subseteq \operatorname{Terms} a \ \&\ \forall b(b \in \operatorname{Dom} f \rightarrow \operatorname{Terms} b \subseteq \operatorname{Dom} f).$$

Let $f = \operatorname{Max} f\, \alpha$. We have α: $\operatorname{Dom} f \leq \operatorname{Terms} a$, $\operatorname{Sup} f \leq a$ and we have $\alpha_f[0]$, so α holds by MAX.

Suppose $\exists d(d \in \operatorname{Terms} a \ \&\ d \notin \operatorname{Dom} f)$. By BLNP there exists a minimal such d. Then there exist b and c such that $d = \tilde{0} \vee (b \in \operatorname{Dom} f$ & $c \in \operatorname{Dom} f$ & $(d = \tilde{\mathrm{S}}b \vee d = \tilde{\mathrm{P}}b \vee d = b\tilde{+}c \vee d = b\tilde{\cdot}c))$.

Suppose $d = \tilde{0}$ and let $f_1 = f \cup \{\langle\tilde{0}, 0\rangle\}$. Then $\alpha_f[f_1]$, so $f_1 \leq f$, a contradiction, and thus $d \neq 0$.

Suppose $d = \tilde{\mathrm{S}}b$ and let $f_1 = f \cup \{\langle\tilde{\mathrm{S}}b, \mathrm{S}f(b)\rangle\}$. Then

$$f_1(\tilde{\mathrm{S}}b) = f(b) + 1 \leq 2^{\operatorname{Ln} b} + 1 \leq 2^{\operatorname{Ln} \tilde{\mathrm{S}}b} \leq \tilde{\mathrm{S}}b,$$

so $\alpha_f[f_1]$ and thus $d \neq \tilde{\mathrm{S}}b$.

Suppose $d = \tilde{\mathrm{P}}b$ and let $f_1 = f \cup \{\langle\tilde{\mathrm{P}}b, \mathrm{P}f(b)\rangle\}$. Then

$$f_1(\tilde{\mathrm{P}}b) = f(b) - 1 \leq 2^{\operatorname{Ln} b} \leq 2^{\operatorname{Ln} \tilde{\mathrm{P}}b} \leq \tilde{\mathrm{P}}b,$$

so $\alpha_f[f_1]$ and thus $d \neq \tilde{\mathrm{P}}b$.

Suppose $d = b\tilde{+}c$ and let $f_1 = f \cup \{\langle b\tilde{+}c, f(b) + f(c)\rangle\}$. Then

$$f_1(b\tilde{+}c) = f(b) + f(c) \leq 2^{\mathrm{Ln}\,b} + 2^{\mathrm{Ln}\,c} \leq 2^{\mathrm{Ln}\,b+\mathrm{Ln}\,c} \leq 2^{\mathrm{Ln}\,(b\tilde{+}c)} \leq b\tilde{+}c,$$

so $\alpha_f[f_1]$ and thus $d \neq b\tilde{+}c$.

Hence $d = b\tilde{\cdot}c$. Let $f_1 = f \cup \{\langle b\tilde{\cdot}c, f(b) \cdot f(c)\rangle\}$. Then

$$f_1(b\tilde{\cdot}c) = f(b) \cdot f(c) \leq 2^{\mathrm{Ln}\,b} \cdot 2^{\mathrm{Ln}\,c} = 2^{\mathrm{Ln}\,b+\mathrm{Ln}\,c} \leq 2^{\mathrm{Ln}\,(b\tilde{\cdot}c)} \leq b\tilde{\cdot}c,$$

so $\alpha_f[f_1]$ and thus $\mathrm{Dom}\, f = \mathrm{Terms}\, a$, and we have $\mathit{scope}_{\exists!f}\,(29.12)$.

Suppose $\mathit{scope}_{\exists!f}\,(29.12)_f[f_1]$, and suppose $\exists d(d \in \mathrm{Terms}\, a \;\;\&\;\; f(d) \neq f_1(d))$. By BLNP there exists a minimal such d, which is impossible. Thus $f = f_1$, thus $\mathit{con}\,(29.12)$, and thus (29.12).

29.13 *Def.* $\mathrm{Valfn}\, a = f \leftrightarrow \mathit{hyp}\,(29.12) \;\;\&\;\; \mathit{scope}_{\exists!f}\,(29.12)$, otherwise $f = 1$.

29.14 *Def.* $\mathrm{Val}\, a = \mathrm{Valfn}\, a(a)$.

29.15 *Def.* A is a variable-free formula of $l \leftrightarrow A$ is variable-free $\;\&\;$ A is a formula of l.

29.16 *Thm.* A is a variable-free formula of $\mathrm{Lang}\,\overline{Q}_0 \;\;\&\;\; B \in \mathrm{Atoms}\, A \rightarrow \exists a \exists b(a$ and b are variable-free terms of $\mathrm{Lang}\,\overline{Q}_0 \;\;\&\;\; B = a\tilde{=}b)$.

Proof. Suppose $\mathit{hyp}\,(29.16)$. Then $\langle B(1)\rangle$ is a predicate symbol, so $\langle B(1)\rangle = \tilde{=}$. Hence there exist a and b such that a and b are variable-free terms of $\mathrm{Lang}\,\overline{Q}_0 \;\;\&\;\; B = a\tilde{=}b$. Thus (29.16).

29.17 *Thm.* A is a variable-free formula of $\mathrm{Lang}\,\overline{Q}_0 \rightarrow \exists! v(v$ is a truth valuation on $A \;\;\&\;\; \forall a \forall b(a\tilde{=}b \in \mathrm{Dom}\, v \rightarrow (v(a\tilde{=}b) = 1 \leftrightarrow \mathrm{Val}\, a = \mathrm{Val}\, b)))$.

Proof. Recall the defining axiom (24.6) for truth valuations. Write α as an abbreviation for $\mathit{scope}_{\exists!v}\,(29.17)$ but with "v is a truth valuation on A" replaced by $\mathit{rhs}\,(24.6)_u[A]$ but with $\mathrm{Dom}\, v = \mathrm{Formulas}\, A$ replaced in it by $\mathrm{Dom}\, v = \mathrm{Formulas}\, A \;\;\&\;\; \forall B(B \in \mathrm{Dom}\, v \rightarrow \mathrm{Formulas}\, B \subseteq \mathrm{Dom}\, v)$. Suppose $\mathit{hyp}\,(29.17)$ and let $v = \mathrm{Max}\, v\, \alpha$. We have α: $\mathrm{Dom}\, v \leq \mathrm{Formulas}\, A$, $\mathrm{Sup}\, v \leq 1$ and we have $\alpha_v[0]$, so α holds by MAX. Suppose

$$\exists B(B \in \mathrm{Formulas}\, A \;\&\; B \notin \mathrm{Dom}\, v).$$

By BLNP there exists a minimal such B. By (29.16) there exist a, b, C, and D such that a and b are variable-free terms of $\mathrm{Lang}\,\overline{Q}_0$ & $(B = a\tilde{=}b \lor (C \in \mathrm{Dom}\,v \;\&\; D \in \mathrm{Dom}\,v \;\&\; (B = \tilde{\neg}C \lor B = C\tilde{\lor}D)))$.

Suppose $B = a\tilde{=}b$. There exists z such that $(z = 1 \leftrightarrow \mathrm{Val}\,a = \mathrm{Val}\,b)$ & $(z = 0 \leftrightarrow \mathrm{Val}\,a \neq \mathrm{Val}\,b)$. Let $v_1 = v \cup \{\langle B, z\rangle\}$. Then $\alpha_v[v_1]$, so $v_1 \leq v$, a contradiction, and thus $B \neq a\tilde{=}b$.

Suppose $B = \tilde{\neg}C$ and let $v_1 = v \cup \{\langle B, 1 - v(C)\rangle\}$. Then $\alpha_v[v_1]$ and thus $B = C\tilde{\lor}D$.

Let $v_1 = v \cup \{\langle B, \mathrm{Max}(v(C), v(D))\rangle\}$. Then $\alpha_v[v_1]$, which again is a contradiction, and thus $scope_{\exists!v}$ (29.17).

Suppose $scope_{\exists!v}$ $(29.17)_v[v_1]$ and suppose

$$\exists B(B \in \mathrm{Formulas}\,A \;\&\; v(B) \neq v_1(B)).$$

By BLNP there exists a minimal such B, which is impossible. Thus $v = v_1$, thus *con* (29.17), and thus (29.17).

29.18 *Def.* $\mathrm{Truthfn}\,A = v \leftrightarrow$ *hyp* (29.17) & $scope_{\exists!v}$ (29.17), otherwise $v = 1$.

29.19 *Def.* $\mathrm{Truth}\,A = \mathrm{Truthfn}\,A(A)$.

29.20 *Thm.* A is a variable-free formula of $\mathrm{Lang}\,\overline{Q}_0$ & (B is an identity axiom $\lor$ B is an equality axiom $\lor$ $B \in \mathrm{Ax}\,\overline{Q}_0$) & A is an instance of B $\rightarrow \mathrm{Truth}\,A = 1$.

Proof. Suppose *hyp* (29.20), and let $v = \mathrm{Truthfn}\,A$.

Suppose 1: B is an identity axiom. Then there exists a such that a is a variable-free term of $\mathrm{Lang}\,\overline{Q}_0$ & $A = a\tilde{=}a$. Then $v(A) = 1$ since $\mathrm{Val}\,a = \mathrm{Val}\,a$. Thus (1) $\rightarrow$ *con* (29.20).

Suppose 2: B is an equality axiom. Then there exist a_1, b_1, a_2, and b_2 such that a_1, b_1, a_2, and b_2 are variable-free terms of $\mathrm{Lang}\,\overline{Q}_0$ & $(A = \tilde{0}\tilde{=}\tilde{0} \lor A = a_1\tilde{=}b_1\tilde{\rightarrow}\tilde{\mathrm{S}}a_1\tilde{=}\tilde{\mathrm{S}}b_1 \lor A = a_1\tilde{=}b_1\tilde{\rightarrow}\tilde{\mathrm{P}}a_1\tilde{=}\tilde{\mathrm{P}}b_1 \lor A = a_1\tilde{=}b_1\tilde{\rightarrow}a_2\tilde{=}b_2\tilde{\rightarrow}a_1\tilde{+}a_2\tilde{=}b_1\tilde{+}b_2 \lor A = a_1\tilde{=}b_1\tilde{\rightarrow}a_2\tilde{=}b_2\tilde{\rightarrow}a_1\tilde{\cdot}a_2\tilde{=}b_1\tilde{\cdot}b_2 \lor A = a_1\tilde{=}b_1\tilde{\rightarrow}a_2\tilde{=}b_2\tilde{\rightarrow}a_1\tilde{=}a_2\tilde{\rightarrow}b_1\tilde{=}b_2)$. Then $v(A) = 1$ and thus (2) $\rightarrow$ *con* (29.20).

Suppose 3: $B \in \mathrm{Ax}\,\overline{Q}_0$. Then $v(A) = 1$ and thus (3) $\rightarrow$ *con* (29.20). Thus (29.20).

29.21 *Def.* A is an open formula $\leftrightarrow$ A is a formula & $\neg(\exists$ occurs in $A)$.

29.22 *Def.* t is an open theory $\leftrightarrow$ t is a theory & $\mathrm{Ax}\,t$ is a set of open formulas.

29.23 *Def.* $\mathrm{Neg}\,A' = \{\langle i, \tilde{\neg} A'(i)\rangle : i \in \mathrm{Dom}\,A'\}$.

29.24 *Def!* t is tautologically consistent $\leftrightarrow$ t is an open theory & $\neg\exists A'(A'$ is a sequence of formulas of $\mathrm{Lang}\,t$ & $\mathrm{Disj}\,\mathrm{Neg}\,A'$ is a tautology & $\forall A(A \in \mathrm{Ran}\,A' \rightarrow \exists B((B$ is an identity axiom $\vee$ B is an equality axiom $\vee$ $B \in \mathrm{Ax}\,t)$ & A is an instance of $B)))$.

29.25 *Def.* $\mathrm{Sub}_0(A, e) = A_{\mathrm{Free}\,A}[\{\langle i, e\rangle : i \in \mathrm{Dom}\,\mathrm{Free}\,A\}]$.

29.26 *Thm.* $\overline{Q}_0$ is tautologically consistent.

Proof. Suppose $\mathit{scope}_{\exists A'}\,(29.24)_t[\overline{Q}_0]$. Let

$$A'_0 = \{\langle i, \mathrm{Sub}_0(A'(i), \tilde{0})\rangle : i \in \mathrm{Dom}\,A'\},$$

and let $v = \mathrm{Truthfn}\,\mathrm{Disj}\,\mathrm{Neg}\,A'_0$. Then $A_0 \in \mathrm{Ran}\,A'_0 \rightarrow v(A_0) = 1$, by (29.20), so $v(\mathrm{Disj}\,\mathrm{Neg}\,A'_0) = 0$ by (24.20). This is a contradiction, and thus (29.26).

Chapter 30

The consistency theorem

We have already discussed the Hilbert-Ackermann Consistency Theorem of [Sh,§4.3] in connection with Assertion 18.1. This theorem is basically an algorithm for eliminating quantifiers from proofs. The first step is to eliminate each use of $\exists$-introduction; this is done by introducing special constants and special axioms. Then the formulas belonging to special constants are eliminated, beginning with special constants of maximal rank. We follow [Sh,§4.3] closely.

30.1 *Def.* $\text{spconstseq}(r', x', C', t) \leftrightarrow r'$ is an injective sequence of constants & x' is a sequence of variables & $\text{Ln}\, r' = \text{Ln}\, x' = \text{Ln}\, C'$ & $\{\langle i, \tilde{\exists} x'(i) C'(i)\rangle : i \in \text{Dom}\, r'\}$ is a sequence of closed formulas & $\forall i \forall j (1 \leq i < j \leq \text{Ln}\, r' \rightarrow \neg(r'(i)$ occurs in $C'(j)))$ & t is a theory & $\text{Lang}\, t \cap \text{Ran}\, r' = 0$ & $\text{Nls}\, C'^{*} \subseteq \text{Lang}\, t \cup \text{Ran}\, r'$ & $\text{Sup}\, r' \leq F_{0, t+\text{Ln}\, r'}$.

30.2 *Def.* $\text{ideqax}(A, t) \leftrightarrow \forall v (v \in \text{Nls}\, A \rightarrow v \in \text{Lang}\, t \vee v$ is a constant$)$ & $\exists B(A$ is an instance of B & $(B$ is an identity axiom $\vee$ B is an equality axiom $\vee$ $B \in \text{Ax}\, t))$.

30.3 *Def.* $\text{belongs}(A, r, r', x', C') \leftrightarrow \exists i (1 \leq i \leq \text{Ln}\, r'$ & $r'(i) = r$ & $(A = \tilde{\exists} x'(i) C'(i) \simeq C'(i)_{x'(i)}[r] \vee \exists a(a$ is a variable-free term & $A = C'(i)_{x'(i)}[a] \simeq \tilde{\exists} x'(i) C'(i))))$.

30.4 *Def.* $\text{Rank}\, A = \text{Deg}(\tilde{\exists}, A)$.

30.5 *Def.* $\text{delta}(A', r', x', C', t, n) \leftrightarrow \text{spconstseq}(r', x', C', t)$ & A' is a sequence of closed formulas & $\text{Nls}\, A'^{*} \subseteq \text{Lang}\, t \cup \text{Ran}\, r'$ & $\forall A(A \in$

$\operatorname{Ran} A' \rightarrow \operatorname{ideqax}(A,t) \vee \exists i(1 \leq i \leq \operatorname{Ln} r' \ \& \ \operatorname{Rank} \tilde{\exists} x'(i)C'(i) \leq n \ \&$ $\operatorname{belongs}(A, r'(i), r', x', C'))).$

30.6 *Def.* $\operatorname{delta}_1(A', r', x', C', t, n,) \leftrightarrow \operatorname{delta}(A', r', x', C', t, n) \ \&$ $\operatorname{Ran}\{\langle i, \tilde{\exists} x'(i)C'(i) \rightrightarrows C'(i)_{x'(i)}\lceil r'(i)\rceil\rangle : i \in \operatorname{Dom} r'\} \subseteq \operatorname{Ran} A'.$

30.7 *Def.* $\operatorname{delta}_0(A', t, n) \leftrightarrow \exists r' \exists x' \exists C' \operatorname{delta}_1(A', r', x', C', t, n).$

We have *rhs* (30.7): $\operatorname{Ln} r' \leq \operatorname{Ln} A'$, $\operatorname{Sup} r' \leq F_{0,t+\operatorname{Ln} A'}$, $\operatorname{Ln} x' \leq \operatorname{Ln} A'$, $\operatorname{Sup} x' \leq \operatorname{Sup} A'$, $\operatorname{Ln} C' \leq \operatorname{Ln} A'$, $\operatorname{Sup} C' \leq \operatorname{Sup} A'$.

30.8 *Thm.* D' is a proof in t of A & A is a closed formula & $\operatorname{Sup} \operatorname{Rank} \circ D' \leq n \ \& \ \varepsilon(2 \uparrow \operatorname{Ln} D') \rightarrow \exists A'(\operatorname{Ln} A' \leq (2 \uparrow \operatorname{Ln} D') - 1 \ \&$ $\operatorname{Sup} \operatorname{Ln} \circ A' \leq 2 \cdot \operatorname{Sup} \operatorname{Ln} \circ D' \ \& \ \operatorname{delta}_0(A', t, n)$ & A is a tautological consequence of A').

(See Lemma 1 of [Sh,§4.3].) *Proof.* Suppose *hyp* (30.8). There exists f such that $\exp(2, 2 \uparrow \operatorname{Ln} D', f)$. Let $k = 2 \uparrow 2 \uparrow \operatorname{Ln} D'$, let

$$r' = \{\langle i, F_{0,t+i}\rangle : i \in \operatorname{Dom} D'\},$$

and write α for

$$\exists r'_0(\operatorname{Ln} r'_0 = \operatorname{Ln} \operatorname{Free} D'(i) \ \& \ \operatorname{Ran} r'_0 \subseteq \operatorname{Ran} r'[i, \operatorname{Ln} D'] \ \& \ A_0 = D'(i)_{\operatorname{Free} D'(i)}\lceil r'_0\rceil).$$

We will prove that

1. $1 \leq i \leq \operatorname{Ln} D' \rightarrow \forall A_0(\alpha \rightarrow \exists A'_0(\operatorname{Ln} A'_0 \leq f(i) - 1 \ \& \ \operatorname{Ln} A'_0 \leq \operatorname{Log} k$ $\& \ \operatorname{Ln} \operatorname{Sup} A'_0 \leq \operatorname{Sup} \operatorname{Ln} \circ A'_0 \leq 2 \cdot \operatorname{Sup} \operatorname{Ln} \circ D' \leq \operatorname{Log}((\operatorname{Maxm} \operatorname{Ln} \circ D')^2) \ \&$ $\operatorname{Sup} \operatorname{Sup} A'_0 \leq \operatorname{Sup} \operatorname{Sup} D' + F_{0,t+i} \ \& \ \operatorname{delta}_0(A'_0, t, n) \ \&$ A_0 is a tautological consequence of A'_0)).

Suppose $\exists i \neg(1)$. By BLNP there exists a minimal such i. Suppose

2. $D'(i) \in \operatorname{Ax} t \vee D'(i)$ is a logical axiom

and suppose α. Let $A'_0 = \langle A_0\rangle$. Thus (1), a contradiction, and thus $\neg(2)$. Suppose

3. $D'(i)$ is a tautological consequence of $D'[1, i-1]$

and suppose α. Then there exists r'_0 such that $scope_{\exists r'_0}\ \alpha$. Recall the defining axiom (29.25) for Sub_0, and let

$$D'_0 = \{\langle j, \operatorname{Sub}_0(D'(j)_{\operatorname{Free} D'(i)}\lceil r'_0\rceil, r'(i))\rangle : j \in \operatorname{Dom} D'[1, i-1]\}.$$

Then A_0 is a tautological consequence of D'_0. Let

$$D''_1 = \{\langle j, \text{Min}\, A'_0\, scope_{\exists A'_0}\, (1)_{i,A'_0} [j, D'_0(j)]\rangle : j \in \text{Dom}\, D'_0\}$$

and let $A'_0 = D''^*_1$. Thus (1), a contradiction, and thus $\neg(3)$. Therefore

$$\exists j(1 \leq j < i \;\;\&\;\; D'(i) \text{ can be inferred from } D'(j) \text{ by } \exists\text{-introduction}),$$

so there exist j, k, B, and C such that

$$1 \leq j < i \;\;\&\;\; x \text{ is a variable} \;\;\&\;\; B \text{ and } C \text{ are formulas} \;\;\& \\ \neg(x \text{ is free in } C) \;\;\&\;\; D'(i) = \tilde{\exists} x B \tilde{\rightarrow} C \;\;\&\;\; D'(j) = B \tilde{\rightarrow} C.$$

Suppose α, so that there exists r'_0 such that $scope_{\exists r'_0}\, \alpha$. Let

$$B_0 = B_{\text{Free}\, D'(i)} \lceil r'_0 \rceil, \\ C_0 = C_{\text{Free}\, D'(i)} \lceil r'_0 \rceil,$$

so that $A_0 = \tilde{\exists} x B_0 \tilde{\rightarrow} C_0$. There exists A'_1 such that

$$scope_{\exists A'_0}\, (1)_{i, A_0, A'_0} [j, B_{0x} \lceil r'(j) \rceil \tilde{\rightarrow} C_0, A'_1].$$

Let $A'_0 = A'_1 * \langle \tilde{\exists} x B_0 \tilde{\rightarrow} B_{0x} \lceil r'(j) \rceil \rangle$. Observe that

$$\text{Ln}\, \tilde{\exists} x B_0 \tilde{\rightarrow} B_{0x} \lceil r'(j) \rceil = 2 \cdot \text{Ln}\, B_0 + 4 \leq 2 \cdot \text{Ln}\, D'(i).$$

Thus (1), a contradiction, and thus (1). Let $i = \text{Ln}\, D'$ and let $A_0 = A$. Since $A = D'(i)$ we have *con* (30.8) by (1), and thus (30.8).

30.9 *Def.* A' is a special sequence $\leftrightarrow$ A' is a sequence of formulas & $\text{Disj Neg}\, A'$ is a tautology.

30.10 *Thm.* e is a constant & D' is a proof in t of $e \tilde{\neq} e$ & $\text{Sup Rank} \circ D' \leq n$ & $\varepsilon(2 \uparrow \text{Ln}\, D') \rightarrow \exists A'(\text{delta}_0(A', t, n)$ & A' is a special sequence & $\text{Ln}\, A' \leq 2 \uparrow D'$ & $\text{Sup Ln} \circ A' \leq 2 \cdot \text{Sup Ln} \circ D')$.

Proof. Suppose *hyp* (30.10). By (30.8) there exists A'_0 such that $\text{delta}_0(A'_0, t, n)$ & $\text{Ln}\, A'_0 \leq (2 \uparrow D') - 1$ & $e \tilde{\neq} e$ is a tautological consequence of A'_0. Let $A' = A'_0 * \langle e \tilde{\neq} e \rangle$. Then *con* (30.10), and thus (30.10).

30.11 *Def.* $\text{Belongs}(A', r', x', C', n) = \{r \in \text{Ran}\, r' : \exists A \exists i(A \in \text{Ran}\, A'$ & $1 \leq i \leq \text{Ln}\, r'$ & $r'(i) = r$ & $\text{belongs}(A, r, r', x', C')$ & $\text{Rank}\, \tilde{\exists} x'(i) C'(i) = n)\}$.

30.12 *Def.* $\text{spdelta}(A', r', x', C', t, n, m, l) \leftrightarrow \text{delta}(A', r', x', C', t, n)$ & $\text{Card Belongs}(A', r', x', C'n) = m$ & $\text{Ln}\, A' \leq l$ & $\text{Ln}\, r' \leq l$ & $\text{Sup Ln} \circ A' \leq l$ & $\text{Sup Ln} \circ C' \leq l$ & t is an open theory & A' is a special sequence.

In [Sh,§4.2], the special constant r for the closed instantiation ∃xA is denoted by the letter c with the subscript ∃xA. This is a beautiful device, which is exploited in the proof of the Consistency Theorem, but a direct arithmetization of it would lead to explosive growth in the size of special constants. Notice that the formula A may itself contain a special constant s. Now r may occur in an expression u without s *occurring* in u, but s will *appear* in u in the subscript of r. The following defining axiom expresses this notion.

30.13 *Def.* $\mathrm{appears}(r,u,r',C') \leftrightarrow \exists j'(\mathrm{Ln}\, j' \le \mathrm{Ln}\, r' \;\;\&\;\; \mathrm{Sup}\, j' \le \mathrm{Sup}\, r' \;\;\&\;\; r'(j'(1)) = r \;\;\&\;\; \forall j(1 \le j < \mathrm{Ln}\, j' \to r'(j'(j))$ occurs in $C'(j'(j+1)))$ & $r'(j'(\mathrm{Ln}\, j'))$ occurs in $u))$.

30.14 *Thm.* u_0 occurs in u & v_0 is a truth valuation on $u_0 \to \exists v(v$ is a truth valuation on u & $v_0 \subseteq v)$.

Proof. Suppose *hyp* (30.14). Write α for

$\mathrm{Dom}\, v_0 \subseteq \mathrm{Dom}\, v \subseteq \mathrm{Formulas}\, u \;\;\&\;\; \mathrm{Sup}\, v \le 1 \;\;\&\;\; \forall B(\tilde{\neg} B \in \mathrm{Dom}\, v \to B \in \mathrm{Dom}\, v \;\;\&\;\; v(\tilde{\neg} B) = 1 - v(B)) \;\;\&\;\; \forall B \forall C(B$ and C are formulas & $B \tilde{\vee} C \in \mathrm{Dom}\, v \to B \in \mathrm{Dom}\, v \;\;\&\;\; C \in \mathrm{Dom}\, v \;\;\&\;\; v(B \tilde{\vee} C) = \mathrm{Max}(v(B), v(C)))$,

and let $v = \mathrm{Max}\, v\, \alpha$. We have $\alpha_v[v_0]$, so α holds by MAX. Suppose

$\exists B(B \in \mathrm{Formulas}\, u \;\&\; B \notin \mathrm{Dom}\, v)$.

By BLNP there is a minimal such B. Suppose 1: B is an atomic formula $\vee\ \langle B(1)\rangle = \tilde{\exists}$, and let $v_1 = v \cup \{\langle B, 0\rangle\}$. Then $\alpha_v[v_1]$, so $v_1 \le v$, a contradiction, and thus $\neg(1)$. Therefore there exist B_1 and C_1 such that $B_1 \in \mathrm{Dom}\, v \;\;\&\;\; C_1 \in \mathrm{Dom}\, v \;\;\&\;\; (B = \tilde{\neg} B_1 \vee B = B_1 \tilde{\vee} C_1)$. Suppose $B = \tilde{\neg} B_1$, and let $v_1 = v \cup \{\langle B, 1 - v(B_1)\rangle\}$. Again we have a contradiction, and thus $B = B_1 \tilde{\vee} C_1$. Let $v_1 = v \cup \{\langle B, \mathrm{Max}(v(B_1), v(C_1))\rangle\}$. This gives a contradiction, and thus $\mathrm{Dom}\, v = \mathrm{Formulas}\, u$. Thus (30.14).

30.15 *Thm.* $\mathrm{spdelta}(A', r', x', C', t, n, m, l) \;\;\&\;\; m \ne 0 \to$
$\exists A_0' \exists r_0' \exists x_0' \exists C_0' (\mathrm{spdelta}(A_0', r_0', x_0', C_0', t, n, m-1, l^2) \;\;\&\;\; \mathrm{Sup}\,(A_0'^{*} * C_0'^{*}) \le \mathrm{Sup}\,(A'^{*} * C'^{*} * r_0'^{*}) \;\;\&\;\; \mathrm{Sup}\, x_0' \le \mathrm{Sup}\, x')$.

(See Lemma 2 of [Sh,§4.3].) *Proof.* Suppose *hyp* (30.15) and suppose

1. $\{\langle i, \tilde{\exists} x'(i) C'(i)\rangle : i \in \mathrm{Dom}\, C'\}$ is injective.

Let

$$
\begin{aligned}
&i_1 = \operatorname{Min} i(1 \le i \le \operatorname{Ln} r' \ \ \& \ \ r'(i) \in \operatorname{Belongs}(A', r', x', C', n)),\\
&r = r'(i_1),\\
&x = x'(i_1),\\
&B = C'(i_1),\\
&A_1' = \operatorname{Enum}\{A \in \operatorname{Ran} A' : \neg \operatorname{belongs}(A, r, r', x, C')\},\\
&A_2' = \operatorname{Enum}\{A \in \operatorname{Ran} A' : \operatorname{belongs}(A, r, r', x', C') \ \ \& \ \ A \ne \tilde{\exists} x B \tilde{\to} B_x\lceil r \rceil\},\\
&a' = \{\langle j, a\rangle : j \in \operatorname{Dom} A_2' \ \ \& \ \ \min_a A_2'(j) = B_a\lceil a \rceil \tilde{\to} \tilde{\exists} x B\}.
\end{aligned}
$$

Then

2. $A_2' = \{\langle j, B_x\lceil a'(j)\rceil \tilde{\to} \tilde{\exists} x B\rangle : j \in \operatorname{Dom} a'\}$

and

3. $\operatorname{Ran}(A_1' * A_2') \subseteq \operatorname{Ran} A' \subseteq \operatorname{Ran}(A_1' * A_2' * \langle \tilde{\exists} x B \tilde{\to} B_x\lceil r \rceil\rangle)$.

We will prove that

4. $1 \le i \le \operatorname{Ln} A_1' \to \neg(\tilde{\exists} x B$ occurs in $A_1'(i))$.

Suppose $\neg(4)$. Then $\neg(A_1'(i)$ is an open formula), and since t is an open theory, $\neg\operatorname{ideqax}(A_1'(i), t)$. Therefore there exists s such that

$$\operatorname{belongs}(A_1'(i), s, r', x', C').$$

There exist y, C, and a such that

y is a variable & C is a formula & a is a variable-free term & $(A_1'(i) = \tilde{\exists} y C \tilde{\to} C_y\lceil s \rceil \vee A_1'(i) = C_y\lceil a \rceil \tilde{\to} \tilde{\exists} y C)$.

But since $\operatorname{Rank} \tilde{\exists} x B = n$, it follows that $\neg(\tilde{\exists} x B$ occurs in $C_y\lceil s \rceil \vee \tilde{\exists} x B$ occurs in $C_y\lceil a \rceil)$. We have $\tilde{\exists} x B \ne \tilde{\exists} y C$ by (1), so $\neg(\tilde{\exists} x B$ occurs in $\tilde{\exists} y C)$. Thus $\neg(4)$ implies (4), and hence (4).

We want to express the notion of replacing each occurrence of $\tilde{\exists} x B$ by $B_x\lceil r \rceil$. We will prove that

5. $\exists g(\operatorname{Dom} g = \operatorname{Formulas} \operatorname{Disj} \operatorname{Neg} A')$ & $\forall A \forall A_1 \forall A_2 \forall z(A \in \operatorname{Dom} g$ & $A_1 \in \operatorname{Dom} g$ & $A_2 \in \operatorname{Dom} g$ & z is a variable $\to$ (i) & $\cdots$ & (vii)), *where*

i. $\mathrm{Ln}\, g(A) \leq \mathrm{Ln}\, A$,
ii. $\mathrm{Sup}\, g(A) \leq \mathrm{Sup}\, A + r$,
iii. A is an atomic formula $\rightarrow g(A) = A$,
iv. $A = \tilde{\exists} x B \rightarrow g(A) = B_x\lceil r \rceil$,
v. $A = \tilde{\neg} A_1 \rightarrow g(A) = \tilde{\neg} g(A_1)$,
vi. $A = A_1 \tilde{\vee} A_2 \rightarrow g(A) = g(A_1) \tilde{\vee} g(A_2)$,
vii. $A = \tilde{\exists} z A_1 \;\&\; A \neq \tilde{\exists} x B \rightarrow g(A) = \tilde{\exists} z g(A_1)$.

Write α as an abbreviation for $scope_{\exists g}$ (5) but with

$$\mathrm{Dom}\, g = \mathrm{Formulas\, Disj\, Neg}\, A'$$

replaced by

$\mathrm{Dom}\, g \subseteq \mathrm{Formulas\, Disj\, Neg}\, A' \;\;\&\;\; \forall A(A \in \mathrm{Dom}\, g \rightarrow \mathrm{Formulas}\, A \subseteq \mathrm{Dom}\, g)$,

and let $g = \mathrm{Max}\, g\, \alpha$. We have $\alpha_g[0]$, so α holds by MAX. Suppose

$$\exists A(A \in \mathrm{Formulas\, Disj\, Neg}\, A' \;\;\&\;\; A \notin \mathrm{Dom}\, g).$$

By BLNP there is a minimal such A, which is impossible. Thus $scope_{\exists g}$ (5), and hence (5).

Let

$$A_3' = \{\langle j, B_x\lceil a'(j) \rceil \tilde{\rightarrow} B_x\lceil r \rceil \rangle : j \in \mathrm{Dom}\, a'\}.$$

We will prove that

6. $A_1' * A_3'$ is a special sequence.

(See (1) of [Sh,§4.3].) Suppose v is a truth valuation on $\mathrm{Disj\, Neg}\, g \circ A'$. Let

$$v_0 = \{\langle A, v(g(A)) \rangle : A \in \mathrm{Dom}\, g\}.$$

Then v_0 is a truth valuation on $\mathrm{Disj\, Neg}\, A'$, so there exists i such that $1 \leq i \leq \mathrm{Ln}\, A' \;\;\&\;\; v_0(\tilde{\neg} A'(i)) = 1$. Therefore $v(\tilde{\neg} g(A'(i))) = 1$, and so $v(\mathrm{Disj\, Neg} \circ A') = 1$. Thus $g \circ A'$ is a special sequence. Recall (3). We have $g \circ A_1' = A_1'$ by (4), $g \circ A_2' = A_3'$ by (2), and $g(\tilde{\exists} x B \tilde{\rightarrow} B_x\lceil r \rceil) = B_x\lceil r \rceil \tilde{\rightarrow} B_x\lceil r \rceil$, so that $g(\tilde{\exists} x B \tilde{\rightarrow} B_x\lceil r \rceil)$ is a tautology. Hence (6).

We want to express the notion of replacing r by $a'(j)$ everywhere it appears, including appearances in subscripts of special constants. Let

$r_1' = \mathrm{Min}\, r_1'(\mathrm{Ln}\, r_1' = \mathrm{Ln}\, r' \cdot \mathrm{Ln}\, a' \leq \mathrm{Log}(r' \# a') \;\;\&\;\; \mathrm{Sup}\, r_1' \leq F_{0,t+l^2} \;\;\&\;\; r_1'$ is an injective sequence of constants $\;\&\;\; \mathrm{Ran}\, r_1' \cap (\mathrm{Lang}\, t \cup \mathrm{Ran}\, r') = 0)$,

$r_1'' = \{\langle j, \{\langle i, r_1'((j-1)\cdot \mathrm{Ln}\, r' + i)\rangle \colon i \in \mathrm{Dom}\, r'\}\rangle \colon j \in \mathrm{Dom}\, a'\}$,
$q = \{i \in \mathrm{Dom}\, r' : \mathrm{appears}(r, C'(i), r', C')\}$,
$r_2'' = \{\langle j, \{\langle i, r_2\rangle : i \in \mathrm{Dom}\, r' \ \ \&\ \ (i \notin q \to r_2 = r'(i))\ \ \&$
$(i \in q \to r_2 = r_1''(j)(i))\}\rangle \colon j \in \mathrm{Dom}\, a'\}$,
$h' = \{\langle j, \{\langle A, \mathrm{Sub}(\mathrm{Ssub}(A, r', r_2''(j)), r, a'(j))\rangle \colon A \in \mathrm{Formulas}\,(A_1' * A_3')^*\}\rangle \colon$
$j \in \mathrm{Dom}\, a'\}$.

(To compute $h'(j)(A)$, first replace all the special constants $r'(i)$ such that r appears in $C'(i)$ by the new special constant $r_1''(j)(i)$ and then substitute $a'(j)$ for each occurrence of r.) Let

$$A_4'' = \{\langle j, \{\langle k, h'(j)(B_x\lceil a'(k)\rceil) \rightleftarrows B_x\lceil a'(j)\rceil\rangle \colon k \in \mathrm{Dom}\, a'\}\rangle \colon j \in \mathrm{Dom}\, a'\}.$$

Suppose $1 \le j \le \mathrm{Ln}\, a'$. Since $A_1' * A_3'$ is a special sequence, $h'(j) \circ (A_1' * A_3')$ is a special sequence. But $h'(j)(B) = B$ since $\neg\,\mathrm{appears}(r, B, x', C')$, so $h'(j)(B_x\lceil r\rceil) = B_x\lceil a'(j)\rceil$. Thus

7. $1 \le j \le \mathrm{Ln}\, a' \to (h'(j) \circ A_1') * A_4''(j)$ is a special sequence.

(See (2) of [Sh,§4.3].)
Let

$$A_0' = A_1' * \{\langle j, h'(j) \circ A_1'\rangle : j \in \mathrm{Dom}\, a'\}^*.$$

We will prove that

8. A_0' is a special sequence.

Suppose $\neg(8)$. Then there exists v_0 such that

v_0 is a truth valuation on $\mathrm{Disj\,Neg}\, A_0'$ & $\forall i \forall j (1 \le i \le \mathrm{Ln}\, r'$ & $1 \le j \le \mathrm{Ln}\, a' \to v_0'(A'(i)) = 1$ & $v_0'(h'(j)(A_1'(i))) = 1)$.

By (30.14) there exists v such that

v is a truth valuation on $\mathrm{Disj\,Neg}\, A_0' * \{\langle j, A_4''(j)^*\rangle \colon j \in \mathrm{Dom}\, a'\}^*$ & $v_0 \subseteq v$.

Suppose $1 \le j \le \mathrm{Ln}\, a'$. By (7),

$$\exists k(1 \le k \le \mathrm{Ln}\, a' \ \ \&\ \ v(h'(j))(B_x\lceil a'(k)\rceil) \rightleftarrows B_x\lceil a'(j)\rceil) = 0.$$

Therefore $v(B_x\lceil a'(j)\rceil) = 0$, and thus

$$1 \le j \le \mathrm{Ln}\, a' \to v(B_x\lceil a'(j)\rceil \rightleftarrows B_x\lceil r\rceil) = 1.$$

This contradicts (6), and thus (8).

Let

$r_0' = \{\langle i, r_3' \rangle : i \in \mathrm{Dom}\, r' \;\; \& \;\; (i \notin q \to r_3' = \langle r'(i) \rangle) \;\; \&$
$(i \in q \to r_3' = \langle r'(i) \rangle * \{\langle j, r_1''(j)(i) \rangle : j \in \mathrm{Dom}\, a'\})\}^*$,
$x_0' = \{\langle i, x_3' \rangle : i \in \mathrm{Dom}\, r' \;\; \& \;\; (i \notin q \to x_3' = \langle x'(i) \rangle) \;\; \&$
$(i \in q \to x_3' = \langle x'(i) \rangle * \{\langle j, x'(i) \rangle : j \in \mathrm{Dom}\, a'\})\}^*$,
$C_0' = \{\langle i, C_3' \rangle : i \in \mathrm{Dom}\, r' \;\; \& \;\; (i \notin q \to C_3' = \langle C'(i) \rangle) \;\; \&$
$(i \in q \to C_3' = \langle C'(i) \rangle * \{\langle j, h'(j)(C'(i)) \rangle : j \in \mathrm{Dom}\, a'\})\}^*$.

Then $\mathrm{spconstseq}(r_0', x_0', C_0', t)$, $\mathrm{Ln}\, A_0' \le l^2$, $\mathrm{Ln}\, r_0' \le l^2$, $\mathrm{Sup\,Ln} \circ A_0' \le l^2$, $\mathrm{Sup}\,(A_0'^* * C_0'^*) \le \mathrm{Sup}\,(A'^* * C_0'^* * r_0')$, and $\mathrm{Sup}\, x_0' \le \mathrm{Sup}\, x'$. We will prove that

9. $A_0 \in \mathrm{Ran}\, A_0' \to \mathrm{ideqax}(A_0, t) \vee \exists s_0(s_0 \in \mathrm{Ran}\, r_0' \;\; \&$ $\mathrm{belongs}(A_0, s_0, r_0', x_0', C_0') \;\; \& \;\; (s_0 \in \mathrm{Belongs}(A_0', r_0', x_0', C_0', n) \to$ $s_0 \in \mathrm{Belongs}(A', r', x', C', n) \;\; \& \;\; s_0 \ne r))$.

Suppose $\neg(9)$. Then $A_0 \notin \mathrm{Ran}\, A_1'$, so there exist j and k such that $1 \le j \le \mathrm{Ln}\, a' \;\; \& \;\; 1 \le k \le \mathrm{Ln}\, a' \;\; \& \;\; A_0 = h'(j)(A_1'(k))$. Let $A = A_1'(k)$. We have $\mathrm{ideqax}(A, t) \to \mathrm{ideqax}(A_0, t)$, so $\neg\mathrm{ideqax}(A, t)$. Therefore there exists i such that $1 \le i \le \mathrm{Ln}\, r' \;\; \& \;\; \mathrm{belongs}(A, r'(i), r', x', C')$. Let $s = r'(i)$, let $y = x'(i)$, and let $C = C'(i)$. There exists a such that

$$A = \tilde{\exists} yC \tilde{\to} C_y[s] \vee A = C_y[a] \tilde{\to} \tilde{\exists} yC.$$

Let $C_0 = h'(j)(C)$, let $s_0 = h'(j)(s)$, and let $a_0 = h'(j)(a)$. Then

$$A_0 = \tilde{\exists} yC_0 \tilde{\to} C_{0y}[s_0] \vee A_0 = C_{0y}[a_0] \tilde{\to} \tilde{\exists} yC_0.$$

We have $s \ne r$, so there exists i_0 such that

$$1 \le i_0 \le \mathrm{Ln}\, r_0' \;\; \& \;\; s_0 = r_0'(i_0) \;\; \& \;\; C_0 = C_0'(i_0) \;\; \& \;\; y = x_0'(i_0).$$

Hence $\mathrm{belongs}(A_0, s_0, r_0', C_0')$. Suppose $s_0 \in \mathrm{Belongs}(A_0', r_0', x_0', C_0', n)$. Then $\mathrm{Rank}\, \tilde{\exists} yC = \mathrm{Rank}\, \tilde{\exists} yC_0 = n$. By definition of i_1 and (30.1),

$$\neg\, \mathrm{appears}(r, s, r', x', C')$$

and so $s_0 = h'(j)(s) = s$. Thus (9), a contradiction, and thus (9). Hence $\mathrm{spdelta}(A_0', r_0', x_0', C_0', t, n, m-1, l^2)$.

Thus (1) $\to$ *con* (30.15). Now let

$i' = \mathrm{Enum}\{i \in \mathrm{Dom}\, r' : \neg\exists j(1 \le j < i \;\; \& \;\; \tilde{\exists} x'(j)C'(j) = \tilde{\exists} x'(i)C'(i))\}$,
$r_5' = r' \circ i'$,

$x'_5 = x' \circ i'$,
$r'_6 = \{\langle i, r'(j)\rangle : i \in \mathrm{Dom}\, r' \;\;\&\;\; \min_j (\tilde{\exists} x'(j) C'(j) = \tilde{\exists} x'(i) C'(i))\}$,
$C'_5 = \{\langle k, \mathrm{Ssub}(C'(i'(k)), r', r'_6)\rangle : k \in \mathrm{Dom}\, i'\}$,
$A'_5 = \{\langle i, \mathrm{Ssub}(A'(i), r', r'_6)\rangle : i \in \mathrm{Dom}\, A'\}$.

Then $\mathrm{spdelta}(A'_5, r'_5, x'_5, C'_5, t, n, m, l)$ & $(1)_{x'C'}[x'_5 C'_5]$. Hence *con* (30.15), and thus (30.15).

30.16 *Thm.* $\mathrm{spdelta}(A', r', x', C', t, n, m, l)$ & $\varepsilon(l)$ & $\varepsilon(2 \uparrow l)$ & $\varepsilon(l \uparrow 2 \uparrow l) \to \exists A'_0 \exists r'_0 \exists x'_0 \exists C'_0 \exists m_0 (\mathrm{spdelta}(A'_0, r'_0, x'_0, C'_0, t, n-1, m_0, l \uparrow 2 \uparrow l)$ & $\mathrm{Sup}\,(A'^*_0 * C'^*_0) \leq \mathrm{Sup}\,(A'^* * C'^* * r'_0)$ & $\mathrm{Sup}\, x'_0 \leq \mathrm{Sup}\, x')$.

Proof. Suppose *hyp* (30.16). Observe that $m \leq l$. Clearly $m = 0 \to$ *con* (30.16), so suppose $m \neq 0$. There exists f such that $\exp(2, l, f)$ and there exists g such that $\exp(l, 2 \uparrow l, g)$. Let $h = g \circ f$ and $k = 2 \uparrow l \uparrow 2 \uparrow l$. We will prove that

1. $1 \leq i < m \to \exists A'_1 \exists r'_1 \exists x'_1 \exists C'_1 (\mathrm{spdelta}(A'_1, r'_1, x'_1, C'_1, t, n, m-i, h(i))$ & $\mathrm{Sup}\,(A'^*_1 * C'^*_1) \leq \mathrm{Sup}\,(A'^* * C'^* * r'_1)$ & $\mathrm{Sup}\, x'_1 \leq \mathrm{Sup}\, x')$.

Let $a = \mathrm{Max}(\mathrm{Sup}\,(A'^* * C'^*), F_{0,t+\mathrm{Log}\, k})$. We have

(1): $\mathrm{Ln}\, A'_1 \leq \mathrm{Log}\, k$, $\mathrm{Ln}\,\mathrm{Sup}\, A'_1 \leq \mathrm{Log}\, k$, $\mathrm{Sup}\,\mathrm{Sup}\, A'_1 \leq a$, $\mathrm{Ln}\, r'_1 \leq \mathrm{Log}\, k$, $\mathrm{Sup}\, r'_1 \leq a$, $\mathrm{Ln}\, x'_1 \leq \mathrm{Log}\, k$, $\mathrm{Sup}\, x'_1 \leq \mathrm{Sup}\, x'$, $\mathrm{Ln}\, C'_1 \leq \mathrm{Log}\, k$, $\mathrm{Ln}\,\mathrm{Sup}\, C'_1 \leq \mathrm{Log}\, k$, $\mathrm{Sup}\,\mathrm{Sup}\, C'_1 \leq a$.

Suppose $\tilde{\exists} i \neg(1)$. By BLNP there exists a minimal such i, which is impossible by (30.15). Thus (1). By $(1)_i[m-1]$ and (30.15) we have *con* (30.16). Thus *con* (30.16), and thus (30.16).

30.17 *Def.* $\mathrm{expcomp}(l, n, f) \leftrightarrow \forall i (i \in \mathrm{Dom}\, f \leftrightarrow i \leq n)$ & $f(0) = l$ & $\forall i (i < n \to \exists a (a \leq f(i+1)$ & $\mathrm{Log}\, a$ is a power of two & $\mathrm{Log}\,\mathrm{Log}\, a = f(i)$ & $f(i+1) = \mathrm{Explog}(f(i), a))$.

30.18 *Thm.* $\mathrm{expcomp}(l, n, f) \leftrightarrow \forall i (i \in \mathrm{Dom}\, f \leftrightarrow i \leq n)$ & $f(0) = l$ & $\forall i (i < n \to \varepsilon(f(i))$ & $\varepsilon(2 \uparrow f(i))$ & $f(i+1) = f(i) \uparrow 2 \uparrow f(i))$.

Proof. From (30.17).

30.19 *Def!* $\sigma_0(l, n) \leftrightarrow \exists f\, \mathrm{expcomp}(l, n, f)$.

30.20 *Thm.* $l_1 \leq l$ & $n_1 \leq n$ & $\sigma_0(l, n) \to \sigma_0(l_1, n_1)$.

Proof. Suppose *hyp* (30.20). There exists f such that $\mathrm{expcomp}(l, n, f)$. Let $g = \{\langle i, f(i)\rangle : i \in \mathrm{Dom}\, f \;\;\&\;\; i \leq n_1\}$. Then $\mathrm{expcomp}(l, n_1, g)$. We will prove that

1. $i \leq n_1 \to \exists h(\text{expcomp}(l_1, i, h) \ \& \ \forall j(j \leq i \to h(j) \leq g(j)))$.

Suppose $\exists i \neg(1)$. By BLNP there exists a minimal such i. Clearly $i \neq 0$. Therefore there exists h_1 such that $scope_{\exists h}\,(1)_{i,h}[i-1, h_1]$. Let $h = h_1 \cup \{\langle i, h_1(i-1) \uparrow 2 \uparrow h_1(i-1)\rangle\}$. Then (1), a contradiction, and thus (1). By $(1)_i[n_1]$ we have $\sigma_0(l_1, n_1)$, and thus (31.20).

30.21 *Thm.* $\sigma_0(l, n) \ \& \ k \leq l \uparrow 2 \uparrow l \to \sigma_0(k, n-1)$.

Proof. Suppose *hyp* (30.21). There exists f such that $\text{expcomp}(l, n, f)$. We will prove that

1. $i \leq n-1 \to \exists h(\text{expcomp}(k, i, h) \ \& \ \forall j(j \leq i \to h(j) \leq f(j+1)))$.

Suppose $\exists i \neg(1)$. By BLNP there exists a minimal such i. Clearly $i \neq 0$. Therefore there exists h_1 such that $scope_{\exists h}\,(1)_{i,h}[i-1, h_1]$. Let $h = h_1 \cup \{\langle i, h_1(i-1) \uparrow 2 \uparrow h_1(i-1)\rangle\}$. Then (1), a contradiction, and thus (1). By $(1)_i[n-1]$ we have $\sigma_0(k, n-1)$, and thus (30.21).

30.22 *Thm.* $\text{spdelta}(A', r', x', C', t, n, m, l) \ \& \ \sigma_0(l, n+1) \to \exists A'_0(A'_0$ is a special sequence $\& \ \forall A_0(A_0 \in \text{Ran}\, A'_0 \to \text{ideqax}(A_0, t)))$.

Proof. Suppose *hyp* (30.22). There exists f such that

$\text{expcomp}(l, n+1, f)$.

Then $\varepsilon(f(n))$. Let $k = 2 \uparrow f(n)$. We will prove that

1. $i \leq n \to \exists A'_1 \exists r'_1 \exists x'_1 \exists C'_1 \exists m_1(\text{spdelta}(A'_1, r'_1, x'_1, C'_1, t, n-i, m_1, f(i))$ $\& \ \text{Sup}\,(A'^*_1 * C'^*_1) \leq \text{Sup}\,(A'^* * C'^* * r'_1) \ \& \ \text{Sup}\, x'_1 \leq \text{Sup}\, x')$.

Let $a = \text{Max}(\text{Sup}\,(A'^* * C'^*), F_{0,t+\text{Log}\,k})$. We have

(1): $\text{Ln}\, A'_1 \leq \text{Log}\, k$, $\text{Ln}\,\text{Sup}\, A'_1 \leq \text{Log}\, k$, $\text{Sup}\,\text{Sup}\, A'_1 \leq a$, $\text{Ln}\, r'_1 \leq \text{Log}\, k$, $\text{Sup}\, r'_1 \leq a$, $\text{Ln}\, x'_1 \leq \text{Log}\, k$, $\text{Sup}\, x'_1 \leq \text{Sup}\, x'$, $\text{Ln}\, C'_1 \leq \text{Log}\, k$, $\text{Ln}\,\text{Sup}\, C'_1 \leq \text{Log}\, k$, $\text{Sup}\,\text{Sup}\, C'_1 \leq a$, $m_1 \leq \text{Log}\, k$.

Suppose $\exists i \neg(1)$. By BLNP there exists a minimal such i, which is impossible by (30.16). Thus (1). By $(1)_i[n]$ we have *con* (30.22), and thus (30.22).

30.23 *Thm.* t is an open theory $\&$ D' is a proof in t of $X_0 \tilde{\neq} X_0$ $\&$ $k = \text{Max}(2 \uparrow (\text{Ln}\, D' + 2), 2 \cdot \text{Sup}\,\text{Ln} \circ D')$ $\&$ $n = \text{Sup}\,\text{Rank} \circ D'$ $\&$ $\sigma_0(k, n+1) \to \neg(t$ is tautologically consistent$)$.

Proof. Suppose *hyp* (30.23). Let $t_0 = \langle \text{Lang}\, t \cup \{\tilde{0}\}, \text{Ax}\, t\rangle$ and let $D'_0 = D' * \langle X_0 \tilde{\neq} X_0\rangle * \langle \tilde{0} \tilde{\neq} \tilde{0}\rangle$. Then D'_0 is a proof in t_0 of $\tilde{0} \tilde{\neq} \tilde{0}$. Observe that $\text{Ln}\, D'_0 =$

$\operatorname{Ln} D' + 2$, so that $2 \uparrow \operatorname{Ln} D'_0 \leq k$. Also, $\operatorname{Sup}\operatorname{Ln} \circ D'_0 = \operatorname{Max}(\operatorname{Sup}\operatorname{Ln} \circ D', 4)$. But since $\operatorname{Ln} D' \geq 1$ we have $k \geq 2^3 = 8 = 2 \cdot 4$, so that $2 \cdot \operatorname{Sup}\operatorname{Ln} \circ D'_0 \leq k$. By (30.10) there exists A'_0 such that

$\operatorname{delta}(A'_0, t_0, n)$ & A'_0 is a special sequence & $\operatorname{Ln} A'_0 \leq k$ & $\operatorname{Sup}\operatorname{Ln} \circ A'_0 \leq k$.

There exists r'_0, x'_0, and C'_0 such that $\operatorname{delta}_1(A'_0, r'_0, x'_0, C'_0, t_0, n)$, so there exists m such that $\operatorname{spdelta}(A'_0, r'_0, x'_0, C'_0, t), n, m, k)$. By (30.22) there exists A'_1 such that A'_1 is a special sequence & $\forall A_1(A_1 \in \operatorname{Ran} A'_1 \rightarrow \operatorname{ideqax}(A_1, t_0))$. Let

$$r'_1 = \operatorname{Enum}\{r_1 \in \operatorname{Nls} A'^{*}_1 : r_1 \notin \operatorname{Lang} t\},$$
$$x'_1 = \{\langle j, X_j \rangle : j \in \operatorname{Dom} r'_1\},$$
$$A' = \{\langle i, \operatorname{Ssub}(A'_1(i), r'_1, x'_1)\rangle : i \in \operatorname{Dom} A'_1\}.$$

By (30.22) we have *con* (30.23), and thus (30.23).

30.24 *Def!* $\sigma(n) \leftrightarrow \sigma_0(n, n)$.

30.25 *Def!* t is σ-consistent $\leftrightarrow$ $\neg\exists D'(D'$ is a proof in t of $X_0 \tilde{\neq} X_0$ & $\sigma(\operatorname{Ln} D'^{*}))$.

30.26 *Thm.* (*consistency theorem*) t is tautologically consistent $\rightarrow$ t is σ-consistent.

Proof. Suppose $\neg$(30.26). There exists D' such that $scope_{\exists D'}$ (30.25). Let $k = \operatorname{Max}(2 \uparrow (\operatorname{Ln} D' + 2), 2 \cdot \operatorname{Sup}\operatorname{Ln} \circ D')$ and let $n = \operatorname{Sup}\operatorname{Rank} \circ D'$. Then $n \leq \operatorname{Ln} \circ D'^{*} - 2$ and $k \leq \operatorname{Ln} D'^{*} \uparrow 2 \uparrow \operatorname{Ln} D'^{*}$, so by (30.20) and (30.21) we have $\sigma_0(k, n+1)$. This contradicts (30.23), and thus (30.26).

30.27 *Thm.* $\overline{Q}_0$ is σ-consistent.

Proof. By (30.26) and (29.26).

Chapter 31

Is exponentiation total?

Why are mathematicians so convinced that exponentiation is total (everywhere defined)? Because they believe in the existence of abstract objects called numbers. What is a number? Originally, sequences of tally marks were used to count things. Then positional notation—the most powerful achievement of mathematics—was invented. Decimals (i.e., numbers written in positional notation) are simply canonical forms for variable-free terms of arithmetic. It has been universally assumed, on the basis of scant evidence, that decimals are the same kind of thing as sequences of tally marks, only expressed in a more practical and efficient notation. This assumption is based on the semantic view of mathematics, in which mathematical expressions, such as decimals and sequences of tally marks, are regarded as denoting abstract objects. But to one who takes a formalist view of mathematics, the subject matter of mathematics is the expressions themselves together with the rules for manipulating them—nothing more. From this point of view, the invention of positional notation was the creation of a new kind of number.

How is it then that we can continue to think of the numbers as being 0, S0, SS0, SSS0, ...? The relativization scheme of Chapter 5 explains this to some extent. But now let us adjoin exponentiation to the symbols of arithmetic. Have we again created a new kind of number? Yes. Let b be a variable-free term of arithmetic. To say that the expression 2^{b} is just another way of expressing the variable-free term of arithmetic, $\mathrm{SS0} \cdot \mathrm{SS0} \cdot \ldots \cdot \mathrm{SS0}$ with b occurrences of SS0, is to assume that b denotes something that is also denoted by a sequence of tally marks. (The notorious three dots are a direct carry-over from tally marks.) The situation is worse for expressions of the form $2 \uparrow 2 \uparrow \mathrm{b}$—then we need to assume that 2^{b} itself denotes something that is also denoted by a sequence of tally marks.

Mathematicians have always operated on the unchallenged assumption that it is possible in principle to express 2^{b} as a numeral by performing the recursively indicated computations. To say that it is possible in principle is to say that the recursion will terminate in a certain number of steps—and this use of the word "number" can only refer to the primitive notion; the steps are things that are counted by a sequence of tally marks. In what number of steps will the recursion terminate? Why, in somewhat more than 2^{b} steps. The circularity in the argument is glaringly obvious.

Although $\varepsilon(n)$ is inductive, one cannot prove $\forall n\, \varepsilon(n)$ in predicative arithmetic. Here are three arguments for this assertion.

First, take a nonstandard element α of a nonstandard model of Peano Arithmetic. Then the set of all elements less than $2 \uparrow \operatorname{Log} \alpha \uparrow k$ for some standard k will be a model of $Q_4'[\exists n \neg \varepsilon(n)]$. This is essentially Theorem 4.3 of Parikh's article *Existence and feasibility in arithmetic* [Pa], which expresses a viewpoint on the foundations of mathematics similar to the one developed here. This argument is model-theoretic; it shows how from a proof of $\forall n\, \varepsilon(n)$ in Q_4' to derive a contradiction in set theory.

Parikh also proves essentially the following (Theorem 4.4 of [Pa]): Let D be a bounded formula such that $\vdash_{Q_4} \exists \mathrm{y} \mathrm{D}$. Then for some bounded term b not containing y, $\vdash_{Q_4} \exists \mathrm{y}(\mathrm{y} \leq \mathrm{b} \;\&\; \mathrm{D})$.

As Sam Buss pointed out to me, this can be used to give a second proof that $\forall n\, \varepsilon(n)$ is not a theorem of predicative arithmetic. Let us use $\vdash$ A to mean that A is a theorem of an extension by definitions of Q_4. Of course, $\vdash \forall n\, \varepsilon(n)$ if and only if $\vdash \varepsilon(n)$, and $\vdash \varepsilon(n) \leftrightarrow \exists f \exp(2, n, f)$. But by Parikh's Theorem 4.4, if $\vdash_{Q_4} \exists f \exp(2, n, f)$, then there is a bounded term b of Q_4 not containing f such that $\vdash_{Q_4} \exists f(f \leq \mathrm{b} \;\;\&\;\; \exp(2, n, f))$. Since $\vdash_{Q_4} \exp(2, n, f) \rightarrow f(n) \leq f$, we would have $\vdash 2^n \leq \mathrm{b}$, which is impossible.

Parikh's proof of Theorem 4.4 is finitary but impredicative. It uses Herbrand's Theorem, and the bound on b is superexponential in the number of occurrences of symbols in the proof of $\exists \mathrm{y} \mathrm{D}$. The second argument shows how from a proof of $\forall n\, \varepsilon(n)$ in Q_4' to derive a contradiction in a theory containing Herbrand's Theorem.

In the definition in Chapter 7 of a bounded extension we required, for each defining axiom of a function symbol with right hand side D, that ∃yD be of bounded form. Since we have the existence condition it would suffice, by Parikh's Theorem 4.4, to require only that D be of bounded form. But in the absence of a predicative proof of Theorem 4.4 it would be unwise to make only this weaker requirement. As we have developed predicative arithmetic, the proof that it is locally interpretable in Q can be arithmetized within the theory itself.

The third argument, which will be outlined now, shows how from a proof of $\forall n\, \varepsilon(n)$ in predicative arithmetic to produce a contradiction in predicative arithmetic itself.

31.1 *Thm.* $\forall n\, \varepsilon(n) \rightarrow ind_n\, \sigma(n)$.

Proof. Clearly $\sigma(0)$. Suppose $\forall n\, \varepsilon(n)$ and suppose $\sigma(n)$. Then there exists f such that expcomp(n, n, f). Let

$$g = f \cup \{\langle n+1, f(n) \uparrow 2 \uparrow f(n)\rangle\}.$$

Then, since $\forall n\, \varepsilon(n)$, we have expcomp$(n, n+1, g)$. Let

$$h = g \cup \{\langle n+2, g(n+1) \uparrow 2 \uparrow g(n+1)\rangle\}.$$

Then expcomp$(n, n+2, h)$, and so $\sigma_0(n, n+2)$. But it follows from (30.21) that $\sigma_0(n+1, n+1)$, i.e. $\sigma(n+1)$. Thus $ind_n\, \sigma(n)$, and thus (31.1). □

Suppose $\vdash \forall n\, \varepsilon(n)$. Then $\vdash ind_n\, \sigma(n)$. Then (30.27) says that $\overline{Q}_0$ is inductively consistent, which for many purposes is as good as full formal consistency. We exploit this as follows. First we arithmetize Q_4. Since it has an axiom scheme, we do this by introducing "t is a $\overline{Q}_4$-theory" to express the notion of an extension by definitions (with an inductive restriction on the number of new symbols) of a finitely axiomatized portion of Q_4. In Chapters 23–28 we developed all the tools necessary (with a fillable gap in the treatment of extensions by definition of a function symbol) to arithmetize the entire development of predicative arithmetic, so that, still under the assumption that $\vdash \forall n\, \varepsilon(n)$, we have $\vdash t$ is a $\overline{Q}_4$-theory $\rightarrow t$ is σ-consistent. The point is that all of our results in the arithmetization of logic took the form of constructing bounded function symbols, which are respected by σ^4 if σ is inductive. In this way (30.27) becomes an inductive self-consistency theorem, and we can employ the familiar reasoning of

Gödel's Second Theorem to derive a contradiction. In doing this, we relativize everything by σ^4; for example, we replace the notion of a proof by a proof such that the number of occurrences of symbols in it satisfies σ^4. Thus if $\vdash \forall n\, \varepsilon(n)$ then predicative arithmetic is inconsistent.

Without the impredicative assumption $\forall n\, \varepsilon(n)$, the formula $\sigma(n)$ is not inductive. Nevertheless, we have $\sigma(0)$ and we have the rule of inference

infer $\sigma(\mathrm{Sb})$ from $\sigma(\mathrm{b})$,

where b is a bounded variable-free term. To see this, suppose $\vdash \sigma(\mathrm{b})$, so that $\vdash \exists f\, \mathrm{expcomp}(\mathrm{b}, \mathrm{b}, f)$. The proof can be relativized by ε_2^4, where ε_2 is as in Chapter 14, so that $\vdash \exists f(\varepsilon_2^4(f) \;\;\&\;\; \mathrm{expcomp}(\mathrm{b}, \mathrm{b}, f))$. Therefore, as in the proof of (31.1), $\vdash \exists g\, \mathrm{expcomp}(\mathrm{b}, \mathrm{Sb}, g)$. Relativizing again by ε_2^4, we have $\vdash \exists g(\varepsilon_2^4(g) \;\;\&\;\; \mathrm{expcomp}(\mathrm{b}, \mathrm{Sb}, g))$, so that $\vdash \exists h\, \mathrm{expcomp}(\mathrm{b}, \mathrm{SSb}, h)$, i.e. $\vdash \sigma_0(\mathrm{b}, \mathrm{SSb})$, and so $\vdash \sigma(\mathrm{Sb})$. Thus if $\vdash \sigma(\mathrm{b})$ then $\vdash \sigma(\mathrm{Sb})$.

We can think of $\sigma(n)$ as a formalization of the notion that n is a genetic number. We do not have the implication $\sigma(n) \rightarrow \sigma(\mathrm{S}n)$ but only the inference, from $\sigma(\mathrm{b})$ infer $\sigma(\mathrm{Sb})$. The distinction expresses the crucial difference between the formal and the genetic; it is the difference between sitting comfortably in camp and saying, "If I have climbed part way up the mountain, then I can always take one more step up the mountain"—and actually climbing a mountain, when each successive step requires an act.

Hence (30.27) is a genetic self-consistency proof for predicative arithmetic. It is a remarkable feature of Robinson's Theory that it can prove its own tautological consistency (29.26). This is made possible by the fact (29.12) that the numbers denoted by terms of the arithmetized theory are bounded by the terms themselves. For a theory T that is strong enough to refer to itself and to prove the full formal consistency theorem, or even an inductive version of the consistency theorem, T cannot prove its tautological consistency without being inconsistent, by Gödel's Second Theorem.

The title of this chapter is not a meaningful question to a nominalist. Instead we may ask, which formula is it more profitable to adjoin to predicative arithmetic, $\forall n\, \varepsilon(n)$ or $\exists n\, \neg\varepsilon(n)$?

Paris and Wilkie (see [PW] and [PW2]) have studied the effect of adjoining $\forall n\, \varepsilon(n)$ and have obtained results relating this to adjoining consistency assumptions. There are a number of interesting questions as to whether various results that can be proved with this hypothesis can be proved, or interpreted, without it. Certainly the theory with $\forall n\, \varepsilon(n)$ adjoined is too weak to serve as a basis for much of contemporary mathematics.

The principal objection to adjoining $\forall n\, \varepsilon(n)$ is that the consistency of the theory is doubtful. One can give a proof of its inductive consistency assuming that superexponentiation is total, or of its full formal consistency assuming that supersuperexponentiation is total. But to prove the consistency of the theories with these additional assumptions, one needs further assumptions yet. It is as if an attorney were to attempt to establish the reliability of a client by bringing in a character witness, and then a character witness to the character witness, and so forth, each one more mafioso than the predecessor. Impredicative finitary reasoning is a residue of Platonism that has been uncritically accepted by the finitists.

As we have seen, it is consistent to adjoin $\exists n\, \neg\varepsilon(n)$, provided only that Q is consistent. Therefore the resulting theory, which may be called Non-Peano Arithmetic, is logically very weak. Nevertheless, it is very strong mathematically. As Hook has shown in his Princeton thesis [Ho], it can be used to develop substantial portions of analysis, by means that will be discussed in the next chapter.

Rather than adjoin $\exists n\, \neg\varepsilon(n)$ to predicative arithmetic, one can try to prove it. This would of course entail the inconsistency of Peano Arithmetic. I have put a lot of effort into this, but so far without success.

Chapter 32

A modified Hilbert program

Hilbert's program was to secure the foundations of classical mathematics by giving a finitary consistency proof for it. This formulation of the program was undoubtedly influenced by his controversy with Brouwer—finitary methods are those (or perhaps a subset of those) that are acccepted by the intuitionists. As far at least as arithmetic is concerned, Hilbert's aim of demonstrating that classical mathematics is no less secure than is intuitionistic mathematics was achieved by Gödel's five page paper [Gö2], published in 1933, in which he gave an interpretation of classical arithmetic within intuitionistic arithmetic. But by then the problem had been utterly transformed by his great paper of 1931. Since finitary methods as commonly understood can be arithmetized, Gödel's Second Theorem doomed the Hilbert program.

From a nominalist understanding of mathematics—from the viewpoint of Hilbert's formalism taken literally—Hilbert's program can be criticized on two counts. First, as I have argued at length, the acceptance of impredicative finitary methods entails a view of mathematics that still contains a semantic element. Second, a convinced formalist should investigate the consistency of classical mathematics as a genuinely open question.

Since the original Hilbert program failed, is it sterile to suggest a program with stricter requirements for acceptable evidence of consistency? Perhaps not; a modification of Hilbert's program appears to be feasible.

The modified program is to build up, parallel to classical mathematics, a demonstrably consistent elementary mathematics such that most results in the core of classical mathematics have an elementary analogue with the same scientific content, and such that the equivalence of the classical to the elementary result is easily provable by classical means.

Here is a candidate, which I will call Q^*, for such an elementary mathematics. To begin with, Q^* should contain predicative arithmetic. With a positive solution to the compatibility problem of Chapter 15, it would be easy to make precise what this means. Lacking this, we must proceed in a more piecemeal way. The theory Q^* is to contain Q_4, the higher # symbols introduced by Hook [Ho], and semibounded replacement as in Chapter 22. Adjoin a unary predicate symbol ϕ and the axioms that say that ϕ respects each bounded function symbol and bounded nonlogical axiom. Adjoin a constant N with the axioms $\varepsilon(N)$, $\varepsilon(2 \uparrow N)$, $\varepsilon(2 \uparrow 2 \uparrow N)$, ... So far, the theory is locally interpretable in Q. Finally, adjoin the axiom $\neg\phi(N)$.

Here is a consistency argument for Q^*. Let us omit the higher # symbols; the conceptual issues are the same. Interpret ϕ as ε^4. Then, as we have seen, $Q_4[\exists n \neg\phi(n)]$ is consistent if Q is. In this theory, introduce a constant N_0 with $\neg\phi(N_0)$ and introduce N by $N = \mathrm{Log}\ldots\mathrm{Log}\, N_0$, with $\nu + 1$ occurrences of Log. Then we cannot prove $\phi(N)$, for if we could, then by $\nu + 1$ successive relativization arguments we could prove $\phi(N_0)$. Therefore it is consistent to adjoin $\neg\phi(N)$, and we have $\varepsilon(N)$, $\varepsilon(2 \uparrow N)$, ..., and $\varepsilon(2 \uparrow \cdots \uparrow 2 \uparrow N)$ with ν occurrences of 2 (and ν is arbitrary).

This is a finitary but impredicative consistency proof for Q^*; the semantic element in the proof enters in the usual proof that Q is consistent, which involves the assumption that superexponentiation is total. A proof of the inductive consistency of Q^* can be based on the assumption that exponentiation is total (so that $\sigma(n)$ is inductive). There are indications that it should be possible to demonstrate the genetic consistency of Q^* with no appeal to semantics at all, that Q^* is truly demonstrably consistent, but such an investigation must await the future.

The formula $\phi(n)$ is similar to the formula "n is standard" in one version [Ne] of nonstandard analysis; the axioms of Q^* asserting that ϕ respects all bounded function symbols and bounded nonlogical axioms have the consequence that if $\vdash_{Q^*} \mathrm{D}$, where D is bounded, then $\vdash_{Q^*} \mathrm{D}^{\phi}$—this is similar to the transfer principle of nonstandard analysis; the axiom $\neg\phi(N)$ is similar to the idealization principle of nonstandard analysis; the use of a many-sorted theory in Hook's thesis [Ho] is perhaps similar to the standardization principle of nonstandard analysis. What we have in Q^* is a grafting of Abraham Robinson's extremely powerful methods of nonstandard analysis (see [Rn]) onto Raphael Robinson's theory.

The theory Q^* is logically very weak but mathematically very strong. The number N such that $\neg\phi(N)$ is in some sense an infinite number, since ϕ is inductive, and yet the axioms $\varepsilon(N)$, $\varepsilon(2 \uparrow N)$, ... allow us to construct the set of all numbers smaller than N and iterated power sets of this set. Once one has an infinite, or in a better terminology an unlimited, natural number, one has infinitesimal rational numbers, and one can introduce real numbers as a new sort of object: a real number is a limited rational number, and two real numbers are equal in case their difference is infinitesimal; see [Ho]. In this way, Hook develops a substantial portion of real and p-adic analysis.

Probability theory is very well adapted to such an approach. Consider the set of all 2^N paths indexed by the rational numbers k/N for $k = 0, \ldots, N$, each of which goes up or down by $1/\sqrt{N}$ at each step. This is an elementary nonstandard model of Brownian motion and, for example, Wiener's theorem that the paths of Brownian motion are almost surely continuous can be given an elementary formulation and proof in this context. Also, the classical form of Wiener's theorem is easily deducible from the elementary form by classical, non-elementary means. This will be discussed in detail in a book [Ne2] that is in preparation.

I am not, of course, suggesting that classical mathematics should be abandoned in favor of an elementary mathematics such as Q^*. Any theory is consistent until proved inconsistent. But I am confident that the modified Hilbert program will repay the efforts of those who find it interesting. The advent of computers has raised the problems of computational complexity to a position of central importance in mathematics. In the world of feasible computations, exponentiation is not total. For results relating questions of computational complexity to syntactical questions in portions of arithmetic, see [Bu] and the references cited there. Nonstandard analysis and computers are recent developments that are transforming mathematics, and the semantical interpretation of mathematics is perhaps unsuitable for both of these developments. I hope that a mathematics shorn of semantical content will prove useful as we explore new terrain.

Bibliography

[Bu] Samuel R. Buss, *Bounded Arithmetic*, Thesis, Princeton University, 1985.

[Fe] S. Feferman, *Arithmetization of metamathematics in a general setting*, Fundamenta Mathematicae 49 (1960), 35–92.

[Gö] K. Gödel, *Über formal unentscheidbare Sätze der Principia Mathematica und verwandter Systeme*, Monatshefte f. Math. u. Physik 38 (1931), 173–198.

[Gö2] —, *Zur intuitionistischen Arithmetik und Zahlentheorie*, Ergebnisse eines math. Koll. 4 (for 1931–32, pub. 1933), 34–38.

[Ho] Julian L. Hook, *A Many-Sorted Approach to Predicative Mathematics*, Thesis, Princeton University, 1983.

[Kl] Stephen Cole Kleene, *Introduction to Metamathematics*, 1952, North-Holland, New York.

[MRT] A. Tarski, A Mostowski, and R. Robinson, *Undecidable Theories*, 1953, Amsterdam.

[Ne] Edward Nelson, *Internal set theory: a new approach to nonstandard analysis*, Bull. Amer. Math. Soc. 83 (1977), 1165-1198.

[Ne2] —, *Radically Elementary Probability Theory*, book in preparation.

[Pa] Rohit Parikh, *Existence and feasibility in arithmetic*, J. Symb. Logic 36 (1971), 494–508.

[PD] J. B. Paris and C. Dimitracopoulos, *A note on the undefinability of cuts*, J. Symb. Logic 48 (1983), 564–569.

[PW] J. B. Paris and A. J. Wilkie, *Δ_0 sets and induction*, in *Open Days in Model Theory and Set Theory*, Proceedings of a conference held in Sept. 1981 at Jadwisin, Poland, Leeds University (1983), 237-248.

[PW2] A. J. Wilkie and J. B. Paris, *On the scheme of induction for bounded formulas*, to appear.

[Pu] Pavel Pudlák, *Some prime elements in the lattice of interpretability types*, Trans. Amer. Math. Soc. 280 (1983), 255–275.

[Rn] Abraham Robinson, *Non-standard Analysis*, Rev. ed. (1974), American Elsevier, New York.

[Ro] R. M. Robinson, *An essentially undecidable axiom system*, Proc. Int. Cong. Math., Cambridge, Mass., 1950, Vol. I, 729–730.

[Sh] Joseph R. Shoenfield, *Mathematical Logic*, 1967, Addison-Wesley, New York.

General index

Index of defining axioms

Library of Congress Cataloging-in-Publication Data

Nelson, Edward, 1932-
Predicative arithmetic.

(Mathematical notes ; 33)
Bibliography: p.
Includes indexes.
1. Constructive mathematics. 2. Arithmetic--
1961- . I. Title. II. Series: Mathematical notes
(Princeton University Press) ; 33.
QA9.56.N45 1986 511.3 86-18730
ISBN 0-691-08455-6 (pbk.)

Edward Nelson is Professor of Mathematics at
Princeton University.

Ingram Content Group UK Ltd.
Milton Keynes UK
UKHW022325160623
423577UK00008B/699